BE A FORCE

HACKING PHYSICS, PERSEVERING LIKE AN OLYMPIAN, AND REALIZING THE ALCHEMIST IN YOU

MICHELLE ROARK

Cover design inspired by artist Mario Duguay
Book design by Anastasia Yagolnik

Publisher's Cataloging-In-Publication Data

ISBN: 978-0-578-50442-1

First Edition

Phi-nomenal, LLC
2150 Curtis Street
Denver, CO 80205

1. Science. 2. Nature. 3. Physics. 4. Flowers. 5. Essential Oils. 6. Michelle Roark. 7. Olympics. 8. Skiing. 9. Energy. 10. Aromatherapy. 11. Wellness. 12. Alchemy. 13. Phia Lab.

www.PhiaLab.com

Dedication

I dedicate this book to everyone who strives to be the best version of themselves, and to their journey of experiencing Heaven on earth by living fully, loving wastefully and being passionate.

Content

Section 3
Performance by Nature

Section 4
Becoming a Force

Introduction

Picture yourself in the start gate at the top of a World Cup mogul course. The pitch is a steep 28 degrees. The finish line where the electronic sensors determine your time is 800 feet down the hill. On the course, there are two jumps, the top air and bottom air, with ramps designed to lift skis 20 to 30 feet into the sky. The best freestyle skiers in the world perform back fulls, back lays, 720's, D-spins and other inverted and vertical creations designed over trampolines and water ramps in the off-season. Your score is based on time and seven judges' opinions on turn technique and acrobatic performance in the air.

In the start area are your coach, the starter and other course technicians. You've been on skis since you were three years old and have devoted your life to the sport. You could be in Finland near the Arctic Circle or sweet home Colorado. You have less than 30 seconds to ski 20 to 25 mph at your peak on terrain that no 4x4 could handle. The wind kicks up and the snow starts to fall on the icy course. "*Skier ready ...*" booms the announcer over the loudspeaker and draws the crowd's attention to the start gate. "*Three, two, one...*" Your moment has arrived. Are you ready?

How would you feel? Nervous? Scared? Prepared? Pumped?

There are few constants across a season or a multi-year career in mogul skiing. Variable factors include the travel itinerary, the course terrain, the blend of snow and ice, weather on competition day, the training meals, coaching, start time, run order, etc. There was only one part of mogul skiing that I could rely on at every competition – pressure. I don't mean the butterfly

> Three, two, one... your moment has arrived. Are you ready?

kind of nervous pressure that one may experience when stepping onto the stage for a speech or musical performance. My feeling of pressure ran deep to my core. Talking was difficult. Tears would often well up in my eyes, and I could find myself too scared to look down the course from the start gate.

Before the competition run, I would sometimes be found in the lodge releasing intense emotions through crying. Competition run training experiences often drove these emotions. On many occasions, I would not be able to complete one top-to-bottom practice run without falling or skiing off the course to one side or the other.

Looking down out of the start gate at the way the steep, icy top air drops away into the clouds would terrify me. This was true in the year that I started my career all the way through to my final season. It never got easier. Every time I pushed out of the start gate, I strived to perform beyond the limits of the run, competition or season before. As the limits extended, the stakes also increased.

However, every time that starter counted me down … "skier ready," I nodded as I had no words. "3, 2, 1." I breathed deep. "Beeeeeeep." I pushed out of the gate, overcame intense emotions and skied flawlessly!

How did I ski my best when it counted the most at every competition? I had over 150 World Cup starts, and I did this miraculous thing 147 times. How did this happen on competition day on the World Cup tour year after year? Looking back, I attribute my long track record of performance to a personal regimen anchored by nature. By harnessing bioenergetics, energy in biological essences found in the botanical world, and combining them in a very specific way I created nutrients for my energy body. Similar to vitamins or supplements designed to benefit our physical body, we can nurture our energy body with what I like to call quantum nutrition.

Quantum nutrition is a compilation of bioenergetics with constructive frequencies combined in phi proportions. These precise combinations can be hyper-customized to one's own signature frequency. In nurturing the energy, body you are better equipped to:

1. Overcome and protect yourself from negative energy.
2. Match the frequency of the reality you want.
3. Attain the state of being in the zone, complete coherence between mind, body and energy.

Start Gate
Olympic Games, Torino Italy

Finish Line
Olympic Games, Torino Italy

After I retired from professional skiing, I became very interested in proving from a scientific perspective how and why the energies around me affected my performances. This translated into two main questions:

1. Why and how did the behavior tactics I developed help me perform at an elite level for so long?
2. How did the energy in biological compounds found in nature elevate my physical performance?

I traveled the world to pursue perfect performance from my start gate at the top of a World Cup mogul course. I set the goal early, persevered through repeated and unexpected challenges, and ultimately, defied the odds. Now, I am on a journey to open minds to these natural performance principles. I want others to be a force in their personal start gate, be it a corporate role, a teaching stage or athletic performance.

> How did the energy in biological compounds elevate my performance?

Teaming up with scientists and engineers

at the Colorado School of Mines, I dove into research. This research helps to bridge the informational gap in understanding the energies around us, and more specifically, the bioenergetics of nature and the effect these energies can have on people interacting with them.

The findings are rooted in four natural laws of physics:

1. The Natural Law of Entrainment
2. The Law of Attraction
3. The Law of Conservation of Energy
4. Energy = Mass x Conversion.

I am excited about our discoveries to date, and humbled by the vast, unexplored territory ahead. Based on my athletic experiences on the slopes and innovative research in the lab, I believe that the unseen energies around us have the potential to improve our well-being and elevate our life experience.

This research could be broad-reaching in medicine, technology, and society. Connected to this work is my belief that energy medicine or quantum nutrition is the way of the future. There are substantial benefits to all those who learn to reconnect with nature as an assistant to cleansing, amplifying and realigning one's energy.

This is my story; it is a fusion of elite athleticism, scientific discovery and the absolute beauty and beneficial energy in nature.

Ultimately, I hope that the principles of physics and quantum nutrition shared in this book will help you – the professional skier, executive, performing artist, teacher, mother, designer, recreational sports enthusiast, or yoga practitioner – to be the brightest and best version of yourself ... to be a force in all you do!

Chapter 1
Five Years Old and Two Goals

Since I was five years old, I wanted to do two things:

1. Compete in the Olympics, and
2. Be a chemical engineer.

I know what you are thinking. What kind of five-year-old knows that she wants to be an Olympian, let alone a chemical engineer?

At the age of five, I watched Olympic figure skating on TV and created a dream of competing in the Olympics. Later that year, my father took me to a natural gas plant in Coalville, Utah. I remember being enclosed in a gigantic white bubble tent, walking around in the mud and being intrigued by all the lights and activity. When I saw a blond lady wearing a hard hat telling all the men what to do, I said, "Daddy, daddy, what does she do?" He responded, "She is a chemical engineer, honey," and I said, "That is what I want to be!" That year, I locked on to these two goals that provided direction and inspiration through the ups and downs.

Did I understand at that time how powerful these early thoughts would become and where they would take me? I am not referring to the superficial dreamy aspect of this mental activity. The power I know today is in the existence of the actual thought and the emotion tied to it. As a child, perhaps the knowledge of this power was, at its peak, slowly losing strength as my thinking was impacted by others over the years.

> The scent of competition, its impact on thought and the ability of thought to evoke the zone are concepts that became my reality.

Did I foresee then that these goals would be so incredibly difficult to accomplish? Was there any awareness in the conscious or subconscious mind for me or others that these two amazingly different goals would synch up to enable the other?

Did I have any idea that my chemical engineering goal would

Michelle, Houston, TX

At the age of five, I watched Olympic figure skating on TV and created a dream of competing in the Olympics.

lead me down the path of refining *essential* oils from botanicals, not the more traditional oil that one extracts from underground? The scent of competition, its impact on thought and the ability of thought to evoke the zone are concepts that became my reality. How were these stars aligned to form the constellation of a new aspiration?

Skating with Mud on my Tights

With the vision of skating in the Olympics spinning in my head, I recall going to my father night after night begging for the opportunity to take lessons in dance, gymnastics or figure skating. I would follow this general plea with "… but I really, really want to be in figure skating."

At the age of five, the breakthrough occurred. My parents took me to a local ice arena and rented me skates. When they returned a few minutes later, they were astonished to see me doing my clumsy version of the "Hamill Camel," jumping and spinning around on the ice with a grin from ear to ear.

At seven years old, we moved from Utah to Westminster, Colorado, and, to my delight, my parents enrolled me in group skating

Natural Gas Processing Plant
When I was five, I toured a gas plant and knew I wanted to be a chemical engineer.

Michelle, Age 9, First Place
My first love was figure skating.

lessons. Within weeks, they let me take the next step by making the investment in private lessons. Despite finishing in last place at my first competition, I was winning medals the following year. For the next several years, my parents' generosity and time commitment to my love of skating was commendable. In pursuit of my dream, I had no difficulty getting up early in second grade. I would rise at 4:30 am, skating before school and returning to the ice arena immediately after school to train again – five days a week, year round.

"Just mention the word 'Olympics' to Michelle, and her smile grows so bright it could light up a room." Ann Margaret Hall

As reported in a Denver newspaper article, "Just mention the word 'Olympics' to Michelle, and her smile grows so bright it could light up a room," said figure skating choreographer, Olympian, and Swedish national champion Ann Margaret Hall.

At the time, ice skating was the third most expensive sport for young athletes. I was grateful for the financial support from my family as resources were limited. With two brothers who were also very involved in sports, I sometimes had to get myself to the ice arena in the morning. My days at that age would often start with a two-mile jog on the canal behind our house to morning practices on the ice. By the time that I would arrive at the arena, I would often have mud on my tights from the run. Other girls would tease me about it, but I did not mind. I would do anything to pursue my dream of skating in the Olympics. However, this "mud on the tights" became a bit of a metaphor for what was to come.

Down but Not Out

Eight years and 55 medals after my skating career began, my dream was shattered. My parents divorced, and my figure skating expenses were too great of a financial burden for a divided family to manage. During this time, my parents, two brothers and I grew unhappy, bitter, and at times mentally and physically abusive. For a variety of reasons, my mother suffered from severe depression. The negative energy of the situation adversely affected all of us.

During this ugly and bitter time, my mother told me one day, "I do not want you!" She took me out of her house and drove me to my father's place. My father was surprised with this behavior and in frustration responded to my mother's demands by saying, "Well, I don't want her either!" However, he recognized that there were no other options and he reluctantly took me in. It was difficult for everyone, and I knew that I needed to escape the bad energy situation to survive. I stayed with my father long enough to get a driver's license, earn money and buy a car. Many years later I recognized how grateful I was for gaining this insight and first-hand experience into the energy I study today. I mention this because I think many readers can relate to a stressful home environment at some point in their lives. How you choose to respond to these challenges can shape your future.

This was my first learning experience in overcoming and protecting myself from bad or negative energy. I was sad and depressed whenever I was around my family. However, this challenge taught me two very important lessons:

1. I never wanted to be 100 percent reliant on anyone for anything again. I was sad about having to give up on skating, and I knew, from that point forward, that no one would have that kind of control over my desires ever again. If I was going to compete in the Olympics, I had to find a way to do it on my own so nobody could take it from me.
2. No one will ever be able to hurt me like that situation had. This felt so freeing, I never was afraid of being let down or pursuing anything difficult moving forward. These early lessons provided the foundation for hurling myself off huge jumps and landing in a field of moguls, starting a business from nothing and pushing myself to think outside the box of standard beliefs.

Chapter 2
Inspired by Bumps

Everyone faces a few bumps and unexpected turns in their own story. How you choose to perceive your situation makes all the difference. When the door shut on skating, I was determined to find a window to crawl through and continue my pursuit. And, I did not care if I got a little "mud on my tights" along the way!

I still wanted to compete in the Olympics and be a chemical engineer. If I could not succeed in skating, I would succeed in something else.

With that resolution to succeed firmly made, I started competing in freestyle skiing that fall. It was a big leap moving from flat ice and steel blades to skis and uneven moguls. However, I owe my father many thanks for putting me on the mountain at a young age.

My father is an avid skier and an incredible athlete. From as early as I can remember, I was inspired by the joy he would experience through play of all kinds. Weekend ski days in Utah and Colorado going back to when I was four and my brother was five, were fun times. The time we spent with our father on those days was mostly in the car going to and from the resort, but I was ok with that. The ski-day routine would normally start and end with some kind of special treat like a donut in the morning and hot chocolate with extra whipped cream at sundown. In those early years, I learned independence. We would arrive at the resort and my father would ski with his friends. He would meet my brother and me for lunch at the lodge and then again at the car at the end of the day. My brother and I would be responsible for getting our gear on and our ski equipment to the lift. We would entertain ourselves with jumps and bumps on the green runs and in the trees and then be on time for lunch and departure.

When one door closes, find a window and crawl through.

Michelle, Age 3
I owe my father many thanks for putting me on the mountain at a young age.

Michelle, Age 6
Skiing with her brother, Park City, UT.

On occasion, he would take us on a skiing road trip. One of the more memorable ski trips was to Jackson Hole, Wyoming. I was nine years old. We got on the lift and went to the top of the mountain. After a short traverse, we were standing on the top of Corbet's Couloir. According to the Jackson Hole resort, it is the most legendary chute in North America. It has also been referred to as "America's scariest ski slope." It is clearly visible from the tram and is often the subject of debate among friends or inside one's own mind – to ski or not to ski it.

This chute serves as a mental proving ground where pride combats fear. The challenge begins with a 10 to 20-foot drop followed by a narrow stretch of snow in between rock outcrops on both sides at a speedy 45-degree pitch. I chose pride and dropped in. On the first attempt that day, I double-ejected from my bindings on the landing and ended up sliding down the entire chute headfirst on my belly at "Mach" speed while onlookers cringed with concern. Despite warnings from Ski Patrol, we immediately jumped back on the lift to try it again. This time I stayed up on my skis, and, as a result, we went from chute to chute. At the top of one gnarly run, a ski patroller said,

"You can't have those kids on this chute!" My father said, "Watch 'em ski, they ski better than you!" The hot chocolate with whipped cream at the lodge bar was extra tasty that afternoon!

My father said, "Watch 'em ski, they ski better than you!"

We all had fun on that day thanks to my father's undeniable love of skiing. We still have fun skiing together. Last season, he skied over 100 days primarily in Winter Park, and is often referred to by friends on the mountain as the Mayor of Mary Jane (the more technical skiing side of the Winter Park resort).

While I knew I could ski for fun, I never competed in moguls until the unexpected exit from skating. After winning many competitions that first season competing in freestyle skiing, it seemed as if I was off to a great start to accomplishing my goal of competing in the Olympics.

The path from winning competitions to competing in the Olympic Games is seldom a straight line for any young athlete. First, it is important to point out that the Olympics take place every four years or once every 1,460 days. Working backward, the road that leads to the U.S. Freestyle Olympic team selection begins with the World Cup circuit. Country-sponsored Freestyle Mogul Skiing teams from around the world compete in approximately 8 to 13 World Cup events each year from December through March. On average over the years, each team of women and each team of men have two to four available Olympic spots to fill with their top-performing skiers. Skiing performance is primarily based on World Cup results. To show your skills on the World Cup circuit, skiers must be part of a designated team and earn what is known inside the sport as World Cup "starts."

For the U.S. Freestyle Ski Team, there are various ways to earn starts. The U.S. Team consisted of seven to 10 skiers assigned to A, B or C categories ranging from high to low in starts and funding. For example, your position on the World Cup tour could be secured by your prior year World Cup ranking. For other skiers, one could improve their chances of gaining a spot on the U.S. Team and earning starts in North America World Cup events by winning Junior Nationals, Nationals and/or the North American (NorAm) tour the prior year. Strong performance by an up-and-coming skier at the North America events could then lead to additional starts at other venues around the world and eventually a World Cup ranking that could solidify a team spot.

Funding for an aspiring Olympian during my years of competition was primarily based on availability and ranged from zero to something short of full support. For other young skiers as well as U.S. Team veteran skiers who lose their U.S. Team status due to poor performance or injury, the path back to a solid spot on the Team leads through regional competitions and the annual U.S. Team Selections event. As I will describe, my journey to the five multi-colored rings looked more like a Jackson Pollock masterpiece than a yellow brick road.

Turns are essential in bump skiing. The physical mechanics of positioning the skis, boots, legs, body weight, hands and head are learned over years of practice on and off the mountain. It begins with keeping your eyes up and looking four to five bumps ahead of your skis. I would feel each mogul with the tips of my skis and, subconsciously absorb and extend over the backside, upper body and hands stable with thoughtless weight shifting and leg-pumping action below. Little did I know at

> One of the most important turns I would make is my turn to independence with nature at my side.

the time that the challenges and opportunities one faces in life align well with mogul skiing.

One of the most important turns I would make is my turn to independence with nature at my side. I realized after that first season that something was still holding me back. I decided to get out of the difficult family environment in Denver to accomplish my goals and retreat to nature. I moved up to the mountains and enrolled myself in Middle Park High School, near Winter Park, 70 miles from family stress in the city. At first, I stayed in a cabin that my father partly owned. When the cabin was sold, I juggled couch surfing, the back seat of my car and tent time in nature.

One might think this must have been a scary and lonely experience for a young girl, but I had a different perspective. In the mountains, I always felt protected by something so much larger than myself, and I never felt entirely alone.

Torino, Italy
Making turns on Olympic mogul run.

Vancouver, Canada
Navigating the bumps.

This new more independent path was not without its own set of real-life bumps in the road. I faced financial challenges, physical injuries, and naysayers – the people with negative beliefs and opinions of my capability. In the remainder of this chapter, I describe these obstacles. In Section Two I take a deeper look back at these obstacles and my path to overcome them through the lens of physics and bioenergetics of nature.

Keep Eyes Up and Over Financial Challenges

I knew education was the key to accomplishing my hard hat engineering goal. At Middle Park High, I maintained honor roll status throughout my time there. However, this took some time away from earning money to support myself and my skiing career. It was a balancing act. I was solely responsible for putting food on my table and covering necessities. I held down jobs at Carver's Bakery, the Silver Screen Cinema, and Club-T tee-shirt shop during my high school years.

Looking back on this part of my life, I think people were both concerned and impressed with my situation. "Most kids in her situation drop out of school and slip through the cracks. She didn't. That's pretty amazing," said my Middle Park science teacher Wayne Benesch when interviewed and quoted in a Denver newspaper.

During my early years of competing in freestyle skiing, I was creative in managing to afford the things that I needed. I skied in my brother's boots (four sizes too big) and put on boyish hand-me-down clothes that did not fit. I borrowed everything else. I basically looked like a short fat boy. This was a big adjustment from my "ice princess" days, where I was accustomed to pink dresses and flashy rhinestones. But this did not bother me because I was not afraid to dive in and get a little "mud on my tights."

My skating skills and training discipline supported a rapid transition to skiing. With numerous victories in my first year of skiing, I was excited about where this path could lead but also prepared for obstacles that might arise. For example, when I discovered that I had qualified for Junior Nationals, my first thought was not about winning the competition. The consuming thought in my mind was, "How will I afford the airline ticket?"

At this time, I started selling everything I had – my bicycle, my beloved clothes, an old racquetball racquet, – in short, anything and everything that I thought could be sold. I was desperate enough to ask my father for money. He had helped before, but this time he was struggling with his own financial situation. Initially, he turned down my request outright. It certainly did not look like a good investment from his perspective. Eventually, my coach convinced him to lend me the money I needed. My father agreed, but with a condition. He offered the loan if I agreed to pay him back with interest. My father definitely taught me the value of the dollar at an early age. I remain grateful to this day. At this time in his life, he would not hesitate to speak his mind and said in so many words, "I don't think you deserve to go to Junior Nationals and I don't think you will do well there anyway." *

Needless to say, I did travel to the Junior National competition and WON! This victory qualified me for an unfunded position on the U.S. Ski Team. As a member of the Team, I benefitted from the additional motivation to improve my finances, cut living costs and generate income.

* Although my father was not all that supportive in the early years he quickly grew to be one of my loudest fans!

At this point, I was 16 years old. I felt that nature was playing a huge role in protecting me from the heavy energy created by the difficult family environment, negative words, and money issues. I would always retreat to nature or more specifically, the mountains to re-energize. It was there that I found support as well as positivity and good energy. Alone in the mountains, I was at peace. No one there condescendingly told me what I was or was not capable of. The only beliefs that found root at this time were my own. I was definitely in a unique situation for a young person; however, it was the foundation for my success.

Alone in the mountains, I was at peace.

Throughout my early years of competing, I needed to be creative with managing my budget. I hitch-hiked around town and sometimes to competitions. Demonstrating an early entrepreneurial spirit, I sold my U.S. Ski Team gear and Christmas presents for income. I would also avoid buying lift tickets by waking early before the ski area opened and hiking up the mountain to train. One time in Whistler, British

U.S. National Forest
Near Rocky Mountain National Park

On the Fraser River, Winter Park
I would always retreat to nature or, more specifically, the mountains to re-energize.

Columbia, I was hiking up in the summertime to train on a glacier when I was met by a family of bears. I could not find a safe way to get around the bears that morning, so I took it as a sign from nature to take a day of rest. I discourage anyone from attempting to hike up ski mountains to avoid buying the lift ticket. This is just my way of telling you that where there is a will you can usually find a way, unless by chance you meet a family of bears!

Often during the summer months, I lived in a tent pitched on National Forest Service land.

With such tight personal finances, the only thing I wanted for my 17th birthday was food. There were a few Thanksgivings during this time when I would go to the local grocery store for free food. Often during the summer months, I lived in a $30 pup tent. I pitched my tent off a windy dirt road in Winter Park. I realized later that squatting on National Forest Service land was illegal.

You might believe that I was naive on some levels and mature beyond my years on others. My tent was fully equipped with pots and pans that I would bang together at night to scare the bears away. For cold or rainy nights, my rusty Jetta was only a short walk away. Luckily, my five-foot frame fit perfectly lengthwise in the back seat, as the seat in that car did not lay down. Again, I felt very protected and nurtured by something much bigger than myself – nature. I can say that in many ways, the energy from my natural surroundings saved me.

In my first full year on the World Cup tour, I finished the season ranked 17th in the world. Unfortunately, I probably ranked #1 in terms of indebtedness to the U.S. Ski Team. The Team loaned me the money for the travel and expenses of the World Cup tour. I was in a financial hole and needed a solution. That spring I moved back into my pup tent and started working 10 to 12 hours a day, waiting tables at Carver's Bakery. When I would get off work at 10 pm, I would take the left-over pastries to bribe the security guard at the local gym to let me in after hours. I knew there were girls from other countries who had finished below me on the overall World Cup ranking, but they were being paid by their federation to train. I figured if I was going to compete against them, I better stick to my workouts, at the very least.

The reality was that although beneficial in some ways, my isolation and living arrangement was not conducive to accomplishing my two life goals in the long run. I needed a better plan. At the end of six months, I had saved enough money to pay back the U.S. Ski Team, but instead, I decided to use the funds as a down payment on a townhouse. I applied for a loan and got turned down several times.

Desperate times sometimes call for creative measures, and I came up with the idea that took my entrepreneurial skills to another level. I managed to get the keys for a townhouse that I was interested in because I knew the listing agent. I would show the place to potential renters as if it was my own. I interviewed these people and then collected their first and last month's rent and security deposit. I used this money on top of what I had saved as a down payment for the purchase. I was still getting rejected from the banks despite my efforts in getting all dressed up, hitch-hiking to Denver and going to bank after bank applying for a loan. Thankfully, my father came through with a small but necessary "gift" right in the nick of time, and I was able to buy the house. The check was in lieu of any financial support for college and was a huge blessing!

I rented all the bedrooms to the lovely people I had interviewed before I had owned the house. I chose to sleep in the open-air loft to maximize the rental income from the bedrooms. The rent exceeded the mortgage, and I paid off my Ski Team debt. I also earned extra cash each month to help pay for school. I learned to do tile, plumbing and electrical work on the house. In three years, I flipped the house for double the purchase price. There were some big bumps in that stretch. But I stayed focused on the goal with my eyes up and leaned on nature for support.

Most people experience financial challenges at some point in their lives. In Section Two, I describe in more detail how specific laws of physics can help people navigate life's "bumps." I will share more detail on how harnessing the right energies from nature can be a tool or assistant along the journey.

Convert Physical Adversity into Opportunity

Unlike many sports, peak performance on skis often occurs on or near the edge of injury. Physical adversity is part of the territory driven by the variability of the terrain as well as the unpredictability of technical maneuvers (i.e., turns, jumps, or landings). With the number of skiers pushing their limits every day on the World Cup circuit, injury is a visible risk that can be mitigated with skill, practice, and proper preparation.

One week before my first World Cup competition in my second competitive ski season, I suffered a bad fall while training on the Winter Park practice course. The result was career threatening – torn ligaments in my left knee. Surgery was the solution followed by months of physical therapy, strength training and eventually skiing. After surgery, I trained with purpose, attended school and continued working to support myself. It was almost a full year before the doctor said I could ski again.

Excited to be skiing and competing again one year later, I made it about halfway down the first competition run of my third season before my

Obstacles are meant to instruct not obstruct.

left knee gave out… again. Unfortunately, the anterior cruciate ligament had not been repaired properly the first time. This was especially difficult to accept after the amount of effort exerted to get back on skis. The new injury caused more severe damage. Ironically, this happened exactly a year to the day of the previous injury on January 7th.

Most people saw this as an end to my skiing career because most competitive skiers retire after their first knee surgery. However, I saw it as just one more bump on the course to reach my dream. At the same time, I was also gaining a greater awareness of how to turn adversity into opportunity. This adversity became an opportunity to focus on my education. From my perspective, this obstacle was not meant to deter me from my dream; it was instruction not obstruction.

This second knee surgery required more than two years of rehab activity and a third surgery to recover fully. The time off from skiing gave me the opportunity to finish high school and apply to college. I was accepted to Colorado School of Mines, my first choice. Mines satisfied two objectives: 1) It was considered one of the top two schools in the U.S. for chemical engineering, and 2) It was in a state which made the cost affordable for me.

It was exciting to think of opening a new educational chapter in my life. However, it was a real struggle financially. I was working three jobs, going to school full-time, and spending the rest of my time following a rigorous rehab program for my injured knee.

While I was waiting for my chance to ski World Cup again, I pursued my chemical engineering degree with a specialty in petroleum and refining and a minor in environmental science. When the money was tight, I took off three semesters to do paid internships. For Bell Atlantic, I worked as a field engineer, designing and managing fiber optic cable projects. On another assignment in California, I was an environmental, health and safety manager at a Johns Manville fiberglass manufacturing plant. It was during my time with Johns Manville where I had the opportunity to wear a hard hat over my blond hair and tell people what to do. Mission accomplished from the perspective of that five-year-old girl who wanted to be a chemical engineer, wearing a hard hat, telling people what to do.

After a long five years from that second injury, including several years of partial World Cup seasons, I managed to re-earn a full-time spot on the U.S. Ski Team. In that first full-time World Cup season, I emerged from the depths of three knee surgeries and rehab to rocket to number one in the world! Winning the Overall World Cup Tour brought along a fully-funded spot on the U.S. Ski Team as well as my first corporate sponsorship from Sprint Corporation. The 2002 Olympics was just two years away and seemed well within reach.

Olympic athletes often attribute their Olympic success to the support of their friends and family. Given my circumstances, I hoped that Olympic success could be attained without the much talked about "support team." Even so, I devoted time to connect with people who seemed to be inclined to offer support. I set out to build a support team.

The U.S. Ski Team is made up of individual skiers. There is no opportunity for a team medal. In this environment, I came to realize over time that if half the people around you are for you, the other half are against you. Imagine attending a competition where your best friend, daughter or brother just completed an amazing run and the only thing standing between him/her and an Olympic-qualifying podium finish, big corporate sponsor and a funded position on the Team, is the performance of the skier in the start gate. It would be tough not to wish the person coming down the run to make a mistake, right?

Those wishes and thoughts carry an energy that can have an effect on others. This is why a professional athlete gets so good at protecting their energy and the energy around them so quickly. If they don't, destructive energy ends up affecting their performance, and they are unable to stay professional for very long. Through my many years on World Cup I saw so many young, promising athletes come up through the ranks only to compete in a few World Cup competitions and then give up. They just had no tools to deal with the heavy energy they would experience at and around these high-level events.

It is hard to say whether the more successful athletes were protecting their energy from the negative thoughts and wishes of others on a conscious or subconscious level. It is clear to me that the most successful athletes definitely figured out how not to be affected by it. This is also why it is important to surround yourself with people you know and trust to have your best interest at heart, in other words, your "support team." A CEO may refer to it as her advisory board. A professor may call it her team of graduate students. Whether we recognize it or not, it feels good to have a "support team."

During the summer before winning the overall World Cup title, I was fortunate to encounter the generosity and kindness of Dr. Edward Manniko. I feel that every athlete should have a Dr. Manniko on "their team." He is a behavioral optometrist who provided me with a new prescription for my contacts. As part of that testing, he detected a physical deficiency in the way my eyes functioned. My deficiency was so bad that when I told him I was in college, he wondered how I could have been accepted into school, let alone to the Colorado School of Mines. When I told him that I was a skier he thought maybe I competed at the club level. He did not see how I could be performing in school or as a world-class athlete based on his assessment of the way my eyes were functioning. Through his generosity and the countless hours donated by his son Mark Manniko, we corrected the problem. The work consisted of many different visual exercises such as playing catch while balancing on a big physioball wearing glasses that distorted my vision in a dark room with a flickering strobe light. I would come to realize the benefit of this training when I skyrocketed to number one in the world the season following this work. This training was invaluable to my success on the World Cup tour and in school.

On January 7, 2000, the same date the previous two knee injuries occurred, I was competing in Deer Valley, Utah, on a course called "Champion." During the training leading up to the competition, I launched off the top air and landed the jump past the target landing area in a hole. The snow was firm and unforgiving that day. Remarkably, I tore nearly all the ligaments and cartilage in my right knee. This injury was so severe that doctors said I would never compete in professional skiing again. I was very fortunate to have access to Dr. Richard Steadman of Vail, the U.S. Team doctor at the time. He was considered the best knee surgeon in the world. Dr. Steadman tried hard to save every shattered piece of cartilage I had left. Being a mogul skier, the cartilage, which is the cushion between bone-on-bone in the knee, is the most critical element for absorption. I will never forget how Dr. Steadman stayed late the night after surgery waiting for me to recover enough to talk. He came into my recovery room and proceeded to inform me of the extensive damage to my knee. He explained that the damage was so severe that he could not fix it all at once and that we would have to do one or maybe two more surgeries. I barely cried on the outside, but the sadness ran deep into my heart.

This was the biggest physical challenge that I faced, and I would have retired if it were not for what happened next. While I was "rehabbing" in the small resort gym in Winter Park, dressed in my usual uniform with sparkles and lipstick, I met someone. He thought I was a tourist from Texas. He was wearing Nike shorts and a matching shirt. I knew that he was not a local. Long story short, he changed my life forever and for the better. His name was Mike, and he was handsome, charming and genuinely supportive and loving. I started to wonder how far I could get in skiing and in life with this new energy that felt so good. Mike became the newest member of my good energy "team."

With every setback, there is a chance to come back. Mike convinced me to return to the regional level and work my way back up the rankings. This is somewhat like asking a professional football player to go back to the high school level and work his way back to the pro football level. It was hard and challenging, but thanks to Dr. Manniko, Mike (a.k.a. Honey), and Dr. Steadman, I was in pursuit again, this time with support. After two more knee surgeries and three years later, I returned to the World Cup competition. The rehabilitation between 2000 and 2003 seasons also allowed me to come within a few classes of completing my chemical engineering degree.

Physical setbacks can test your resolve or lead to early retirement. Except for my closest support team, nearly everyone I knew, including myself, had given up on me getting back to World Cup form after that sixth surgery. But I learned to turn this obstacle like the many before it into an opportunity. Obviously, it did not hurt that my "workplace" was situated in nature and, as I said, my connection with nature is the foundation on which I have built all my life.

If I had a nickel for everybody who has said to me "I could have" or "my son/daughter or friend could have competed in the Olympics but he/she blew their knee," I would have a full piggy bank. I am not sure I have ever heard of a completely smooth journey. Bumps come in all forms along everyone's path. Sometimes these bumps are there to indicate you are on the wrong path. Other times they are there to test your dedication or teach you something along the way. Through navigating my bumpy path, I developed a unique perspective, behavioral tactics and set of performance tools outlined in Section Two that I hope will help others.

Choose Yaysayers Over Naysayers

Naysayers are people with negative beliefs and opinions. I think talking about naysayers is really important. This is because from what I have seen, people do not fully realize the effect a naysayer can have on their energy. Dr. Masaru Emoto, a Japanese scientist, and water researcher, revealed the true nature of water and how thoughts and energy affect the molecular structure of water.

Back Flip, Mt Hood, OR
Six knee surgeries, but always bounced back!

My Office, Winter Park, CO
My "workplace" was situated in nature.

In this research, he concentrated on high-speed photography of thousands of water crystals. One experiment involved taping words to a glass of distilled water, freezing the water and then looking at the frozen water crystals under a microscope. In the pictures below, as shown in *The Healing Power of Water* by Dr. Emoto, you can see the crystals formed after being exposed to positive and negative words.

The crystals from water exposed to positive words formed beautiful snowflake crystals with light emitting from the center. No crystal formed from the water exposed to negative words, it was just a brownish blob.[1] Although I believe what Dr. Emoto was trying to prove with his experiment from a scientific perspective, his work lacked credibility based on research principles. This opened the door for many researchers to shoot holes through his experiments.

Thank you

You Make Me Sick

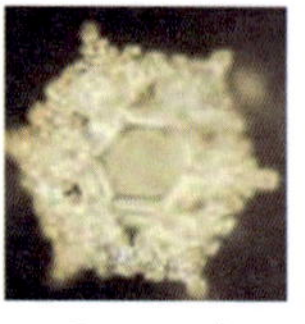

Love and Appreciation

1. Masaru Emoto, The Healing Power of Water (Hay House, California, 2008).

However, Dr. William Tiller, a professor emeritus of materials science and engineering at Stanford University, conducted similar research with substantial scientific experimentation and delivered consistent results. Dr. Tiller did this by performing rigorous experiments on water where he could raise and lower the pH of water by one full point purely by intention or thought.[2] His experiments have been repeated all over the world with profoundly remarkable success.

To give you just one comparison of how remarkable Dr. Tiller's results are, consider the following. If you raise or lower the pH in your blood by just half a point, you are dead on both ends. These results have had immense implications in the medical industry and how we as a society believe in health, wellness, and medicine. We are made up of 70 percent water, and we are actually 99 percent water molecules.[3]

We are now closer than ever to explaining through physics or what some call "new physics" what mystics and yogis have practiced for some time. Thoughts and words carry an energy. As an athlete, you learn early on to protect yourself from the unseen energy around you. You may not

2. William Tiller, Ph.D., Walter Dibble, E., Jr., Ph.D., White Paper I - A Brief Introduction to Intention-Host Device Research, (Tiller Institute, 2009)

3. Gerald Pollack, The Fourth Phase of Water (Ebner & Sons, 2013).

know what you are doing, but you instinctively shield yourself from the energy of those who do not wish you well. From experience and through a deeper understanding of new physics, I can say that if you don't learn to protect yourself from negative energy, it will jeopardize your physical performance.

In my life, there were many naysayers, but I always called upon that childhood saying of "I am rubber, and you are glue, whatever you say bounces off me and sticks to you." Additionally, I invoked my visualization techniques to envision myself floating around in a protected sparkly, iridescent, rainbow-colored bubble. This worked well most of the time. When my father told me that I didn't deserve to go to Junior Nationals, I went and won. When the surgeon told me that I would never compete professionally again, I let this bounce off and got back on the hill. When the head coach of the U.S. Freestyle Ski Team said I was not the elite of the elite, my work outlasted his coaching tenure, and I won the overall World Cup title the year after he retired. Ultimately, I learned that the moments hardest to overcome occur when your closest connections, e.g., your family, do not believe in you.

I felt that my biggest naysayer was my mother. It was an unfortunate situation, but perhaps I would not have found my path without this negative force in the rearview mirror. For example, since I was five, she knew that I wanted to compete in the Olympics. When I was about 10 years old, she went to a psychic with her friend where she inquired about her children's future. She almost beamed as she told me that the psychic had foretold that I would never compete in the Olympics. For a 10-year-old girl, this was devastating news.

At that time, I was without any tools to deal with it, so I suppressed it. Typically, your family is your biggest support system; they are your cheerleaders. Your family is where you turn when you need a pick-me-up. From an energy perspective, you are connected to your family. This is why family members can have the strongest impact as a naysayer. In these unfortunate situations, I suggest complete separation to avoid being adversely influenced by them. Their sheer presence can trigger old belief systems that have been created based on your past experiences with them.

Bruce H. Lipton, Ph.D., is an internationally recognized leader in stem cell biology, bestselling author of *The Biology of Belief* and recipient of the 2009 Goi Peace Award. His research at Stanford University's School of Medicine on stem cells revealed that the living environment surrounding the cell membrane controls the behavior and physiology of the cell.

As part of Lipton's early research in 1967, he experimented with multi-potential cells (i.e., stem cells) and found that one cell placed in a petri dish by itself would divide every 10 or 12 hours. After one week, he would have roughly 50,000 cells in a single dish. More importantly, every cell in the petri dish was genetically identical to every other cell in the dish. These cells came from the same parent. He essentially grew 50,000 genetically identical cells. And then he split the identical cells into three different petri dishes. In each dish he changed the chemicals in the culture medium. So in each dish there was a slightly different chemistry. What happened was that in one dish the cells formed muscle. In a second dish the cells formed bone and in the third dish the cells formed fat cells. Now what controls the fate of the cells? The environment!

Additionally, he has done some remarkable work pointing to the ingrained belief systems imprinted in children by their parents.[4] Dr. Lipton points out that between the ages of zero to seven years, children are downloading the beliefs and patterns of their parents into their subconscious mind, a process that can impact the rest of their lives. As a result, if your parent says you are abundant and capable of great things, your life follows this pattern; but if they say you are worthless, that is what you become unless you learn to reprogram the subconscious mind.

"Between the ages of zero to seven, children are downloading the beliefs and patterns of their parents which then shapes the rest of their life." Dr. Bruce Lipton

Again, I always found nature as a safe retreat to reenergize, reprogram and feel supported whenever other peoples' beliefs, words or actions were in contradiction to the reality I wanted. This is why being isolated in the mountains away from the

4. Bruce Lipton, The Honeymoon Effect (Hay House, Inc, 2013).

negative thoughts and opinions of a struggling family was beneficial for me. I've seen people who remain in that type of situation have a very different outcome.

Imagine the impact of a life full of positive words and thoughts surrounded by nature and compare that to a life full of negative words and thoughts living in a concrete jungle. Do you know someone who seems to have an ability to turn everything she touches into gold? Do you think that person is incredibly blessed by a force outside of themselves, or is it more probable that the person consciously or subconsciously knows the power of words and thoughts?

In some mystical circles, thought is considered the most powerful force in the universe. While this belief may lack scientific proof, I was more than willing to try it.

This concept was evident in my personal experiences, not just once but over and over again.

Many doctors told me I would never ski professionally again, but I competed my last seven years on World Cup injury-free. I did this by retreating to nature and making a conscious effort to choose my path and not letting the beliefs or words of others derail me. This system worked well in most cases. However, I found in situations where the negative words were coming from my own family, more effort was required. I had learned to suppress and forget the negative opinions of my mother and her psychic's prediction since at the time I was not equipped to handle it. It was not until the predicted situation happened that it resurfaced.

In 2002 I was fully expecting to compete in the Salt Lake City Olympic Games, but, unfortunately, I was injured. I ended up commentating on the local television station. It was at this moment that I remembered the entire set of words dating back to when I was 10 years old: "I would never compete in the Olympics, but do something with television." Here I was on television, but not competing.

Once this mental obstacle resurfaced, I used the tools I had become familiar with to clear and reprogram the belief. I will elaborate on these tools in Section Two. As a result, I was able to eliminate this blockage that I was not fully aware of and make room for my dreams to be realized.

Returning to World Cup competition after completely clearing out old beliefs that were not serving me and acquiring a real support team

Keystone, CO
Photoshoot for sponsor Tommy Hilfiger

was a whole new positive experience. Mike would come to all my World Cup competitions no matter where they were held. He worked as a consultant mostly in New York, so he was accustomed to traveling. We typically had about 13 World Cups per season and traveled to different countries nearly every weekend. Mike would work all week in New York then fly to wherever I was on the weekends. This may sound luxurious or exciting, but it was typically a red-eye flight to Europe, then six or so hours via bus, train then taxi to a remote location. Once there, he had two jet-lagged days to deal with the stressed-out me before flying back to New York and doing it all again. For seven years he never missed a World Cup. Once he had his own knee surgery and just five days after that, he hiked up the ski hill on crutches to watch and support me in my competition.

I had always heard from former Olympic athletes that they could have never done it without the support of their family or "support team." For the first half of my professional career, I had searched for a path to the Olympics without a solid support team. For the second half, I found a true support system and learned to let go of subconscious beliefs. This made a huge difference along my path to the Olympics.

I won the silver medal at the 2003 World Championships. As luck would have it, it was on the same Deer Valley course that I had gone down on three years prior.

For the next two years following World Championships I competed on the full World Cup Circuit with one goal in mind: stay healthy! The plan was to maintain my A-Team status (i.e., a funded position with World Cup starts for the next season) and come on strong in the second half of 2005 to gain momentum that would carry me to the 2006 Olympics.

In the second half of the 2005 season, I did just that by winning the Deer Valley World Cup, the same course that took me out in 2000, and placing third at the Lake Placid World Cup. This put me in a great position leading into the Olympic trials the following ski season. The Olympic trials were a series of five World Cup competitions where a podium finish was required to be in the running for one of the two to three Olympic spots available. In these trials, I placed first, first, third, fifth and 11th. My three podium finishes were more than enough to secure my spot. Woohoo! Mission accomplished!

My own physical experience is proof of how the energy around us can affect our physical bodies and performances. Have you ever heard the expressions, "it's just words" or "actions speak louder than words?" I disagree with this. Words contain energy and energy has an effect on us. I plan to show you how we all have an energy body and how that body can develop knots similar to the development of a knot in a sore muscle. Only these knots are not from physical exertion but from emotional exertion. These knots can occur from negative energy through words and thoughts. If these tangles are not properly worked out they get in the way of our desires. Surrounding yourself with a support team is a good way to shield yourself from the negative energy of others, but there are other effective tactics and tools that can also be used.

Chapter 3
The Scent of Competition

In 2005 I had the opportunity to work with the U.S. Ski Team sports psychologist for the first time. She said that I needed to make sure I covered ALL of my senses when I visualize skiing perfectly. In other words, she wanted me to make sure I could see it, hear it, feel it, smell it and taste it. I had no problem with any of these except for the sense of smell. I had no idea what it "smelled" like to ski well. So, I thought I would find a fragrance and wear it just for skiing. It would be my "scent of competition!" However, I was highly dissatisfied with commercial fragrances, and I tried many.

I had no idea what it "smelled" like to ski well.

During my extensive international travel, I had ample time to kill trying fragrances at the airport duty-free shops during layovers and flight delays. They would often give me a headache, be gone within an hour, and, sometimes, they would burn my nose. Determined to find my scent of competition, I kept trying different fragrances. After a while I started looking into what was going into the fragrances I was trying and found them to be quite toxic. In fact, it seemed walking into the cosmetic department of a store could be more toxic for you in terms of what you are smelling than an auto mechanics garage where you have oil, gasoline, and formaldehyde.

This is unfortunate when you consider how many people are spraying this on themselves every day, sometimes multiple times a day, without sparing a thought to the toxins that it contains. More importantly, skin, our largest organ, is not designed to process out toxins like the liver does from what we consume. This made me think something needed to change.

Back Flip and Roses, Winter Park, CO

I wanted to create something that would be empowering and pleasing to the wearer, something that would be my scent for the Olympic Games.

Creating a Scent to Empower

I decided I would apply my background in chemical engineering and make my own scent – my personal "scent of competition" – that would empower and energize me. First, I bought every perfume book I could find and took perfumery courses based in Grasse, Thailand and New York. I started with an all-natural carrier base. No ethanol alcohol or synthetic preservatives were necessary. This approach would also enable one drop to last about six hours on my body. I needed a scent to stick with me through the competition day.

I wanted to create something that would be empowering and pleasing to the wearer, something that would be my scent for the Olympic Games. I had come so close to the Olympics before and had never made it. I decided to expand my quest for technical perfection on the hill to include performance drivers off the hill. On this quest, no stone would be left unturned.

I did this, first in 2005, by harnessing the bioenergetics of nature through the use of pure, natural, biological essences of plants. I also blended these essences using a proprietary "Phi" formulation system.* This was a pivotal event

*Phi is explained in greater detail in chapter 5

in my professional career and my life. I was starting to understand firsthand the incredible benefits contained in plant, flower and herb essences that would carry me through five more years on the World Cup circuit injury-free.

During these five years, I had more success than all the previous 11 years combined. While I did not consciously understand it at the time, the energy in the pure essences was helping me release old, destructive thought patterns and belief systems. These negative thoughts were tangled up in my energy body. I was not consciously aware of them because they had been suppressed for such a long time.

Signature Frequency

Significant gains over the past two decades have been made in understanding our sense of smell. Largely, this research has come out of the Monell Institute, the world's only independent, nonprofit scientific institute dedicated to research on taste and smell.

In the book The Secret of Scent, Luca Turin explains how smell has historically been the least understood but also the most important sense. The generally accepted premise used to be that it is the size and shape of the molecules in a scent that fit different receptors in the nose, which then signals the smell to the brain. Today, it is understood that it is not just the size and shape of the molecule but also the vibration that then indicates the smell to the brain.[5]

An earlier version of this concept came from Turin's 1996 vibrational theory of olfactory reception. The theory is essentially vibrational in nature, looking not at the shape of a molecule but its vibrations, which, like resonant frequencies in a bridge, are a function of the structure.[6] At the same time, research was pointing to the discovery that our smell, the biological odor prints we emit, is more unique to us than our fingerprint.[7]

Gary Beauchamp, a behavioral biologist at the Monell Chemical Senses Center concluded in a study published in the October 31, 2008, issue of the journal PLOS One, that your body has a signature odor, just as your fingers have unique prints; and second, that "Eau d'you" remains even if things in your environment

5. Luca Turin, The Secret of Scent (Ecco, 2007).

6.Joseph Kaye, Symbolic Olfactory Display (Massachusetts Institute of Technology, May 2001).

7. Richard Gerber M.D., Vibrational Medicine (Bear & Company, Vermont, 2001).

> Our vibration or frequency is more unique to us than our fingerprint.

change. If smell is a vibration, then it follows that our vibration or frequency is more unique to us than our fingerprint.

While the frequency or smell that we emit is known to change based on emotions, thoughts, activities, and things we eat, this new research indicates that we all have a "signature" scent and thus a "signature" frequency that is unique to everyone and does not easily change. If we all have a unique signature frequency, then we conclude that we need to look at what we call in physics constructive and destructive frequencies.

Figure 1a shows two waves of equal amplitude interfering constructively. The resulting amplitude is twice as large as the amplitude of the individual wave. For 20 waves of the same amplitude interfering constructively, the resulting amplitude is 20 times larger than the amplitude of one wave. In other words constructive

Figure 1

a) Constructive Interference b) Destructive Interference

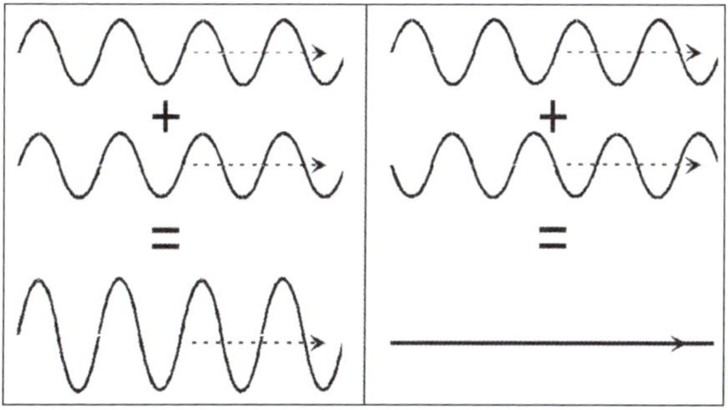

interference can produce a significant increase in amplitude or strength.

On the other hand, destructive interference occurs when waves come together in such a way that they cancel each other out, as seen in Figure 1b. These constructive and destructive frequencies explain why someone may sound like a perfect match for you on paper, but when you meet there is no real connection (i.e., signature frequencies are destructive and cancel each other out). Likewise, someone else may not sound like your type on paper, but when you meet, it is love at first sight (signature frequencies are constructive and amplify each other).

Now picture yourself at the elite level of anything: academia, work or athletics, where one-fourth of a second is the difference between winning and losing, money and no money, failure or success. Wouldn't you want to surround yourself with people of similar frequencies? However, this is often not possible. Similar to colleagues at work or members on a ski team, different people are often brought together by circumstance, not choice. We trained, had meals, traveled and basically spent more time with our teammates than our own family. If there was a frequency in the group not resonating with the rest the result was felt all around.

My thought was if we all have a signature frequency, how can I strengthen it and or nurture it. In doing so perhaps the energy of others, not constructive to mine, would not have an effect. As I thought about it, I realized I was spending many hours strengthening my physical body in the gym. I also took supplements and tried to eat healthy to nourish my physical body, but what was I doing to nourish my energy body? Ultimately, what I found was the very high vibrations in the biological essences I used to create my custom "scent of competition" was acting as quantum nutrition for my energy body. Harnessing these energetic essences and combining them in a very specific way was just the right kind of nourishment my energy body needed. This was reflected to me in my performances on World Cup and my health and wellness in general. Exactly how and why I arrived at this important conclusion is discussed in detail in the research chapter.

Chapter 4
Natural Laws of Physics

The laws of physics are considered fundamental. Some are simple and easy to understand. Others may be harder to absorb because they are less visible in the real world. My seven-year-old is learning that hot chocolate becomes cool if left on the counter. My three-year-old is experiencing the steady force of gravity on the bunny slope. Other laws are more difficult to grasp like some of the new physics (i.e., everything being connected on a quantum level, thoughts creating your reality) and areas like epigenetics (i.e., cells or even your DNA being influenced by the environment).

Extensive experience in World Cup competition combined with rigorous study and independent research in the hard sciences has put me in a unique position. I can pull from real-world experiences to back up repeated findings in the lab. In the next four sections I will discuss four natural laws of physics:

1. The Law of Entrainment
2. The Law of Attraction
3. The Law of Conservation of Energy
4. Energy = Mass x Conversion

I will do my best to explain these natural laws from a scientific perspective, but also through real-life experiences.

The Law Of Entrainment

The Law of Entrainment or Entanglement is a universal law that states any two vibrating bodies will entrain or absorb the energy of the other if exposed to each other long enough. This could be a good thing if the people you are spending most of your time with have a constructive frequency to your own and thus amplify your frequency. A good example of this could be seen in the Finnish Freestyle Mogul

Team where, at one point, all team members were either related somehow or longtime friends. During this period, they often swept the podium at World Cup competitions.

Another example was the U.S. Freestyle Aerial Team in the early 2000s. They were all best friends and still are today. During their World Cup years, they often swept the podium. On the flip side, if just one member had a destructive frequency to the group, it would throw off everyone, similar to how one rotten apple spoils the barrel. This is also why a home or work environment filled with negative words, thoughts and feelings is unhealthy for everyone.

Aside from the energy in the thoughts and opinions of others as mentioned in Chapter Two, there is also entrainment with possible destructive signature frequencies of the people with whom you must associate. Consciously or subconsciously, athletes competing on the World Cup circuit found ways of dealing with this. Some would go for long runs, mask it with heavy partying or opt for some meditation. What I found refuge in was the vibration of nature or more specifically, the vibration in nature's botanicals.

If I could match my home frequency in a constructive way to the vibration in these botanicals, as I did with my "scent of competition," it would amplify, cleanse and realign my energy, counteracting the effect of destructive frequency. I found this to play an important role in illuminating my energy thereby improving my health, wellness and my physical performance. At the time, I was doing this on a subconscious level. Later, I was able to recognize it and prove it through research at Phia Lab.

The first step is to recognize that there is energy all around us. We are energetic beings with our own signature frequency. Swami Vivekananda said it best: "To understand the true nature of things, one must think [of it] in terms of energy, frequency, and vibration." Through the natural Law of Entrainment, the energies around us can affect our energy and thus, our physical body in positive and negative ways. Once you realize this, you can find ways to protect your energy and illuminate it.

First, try to surround yourself with people of constructive frequency who have nothing but positive thoughts and kind words for you. Shield yourself from destructive frequencies by avoiding certain people and thereby protecting yourself from the harsh words and thoughts of others. You can also do this by visualization techniques of a protective layer of light, iridescent bubble or suit of armor surrounding you.

My favorite is utilizing the high vibrations found in nature or, more specifically, in the biological essences of the botanical world. I found that when combined in the right way, these essences can effectively cleanse, amplify and align your energy body, so you are the stronger system. As a result, and according to the natural Law of Entanglement, the weaker systems entrain to you.

Illuminate your energy body with constructive frequencies found in nature and be the stronger system.

An excellent example of this principle is if you can picture a fast-moving river rapidly flowing along its path. Now pick up a small piece of dried wood and throw it in the river. The wood does not try to change the direction of the flowing water nor does it speed up or slow down the running water because it is the weaker system. Instead, the wood entrains with the energy of the flowing river. I always tried to be like the river and be the stronger system by illuminating my energy body with constructive frequencies found in nature, or what I like to call quantum nutrition, so others had to entrain with me instead of vice versa. This was a huge contributor to my long-term success.

In the next section I will take this concept of energy and illumination a step further. I will explain why protecting your energy is important, especially for attaining aspirations, being more effective day-to-day and having more energy at the end of your day.

The Natural Law of Attraction

In skiing, I applied a scientific approach to assessing and re-assessing my technique. I sought out the best coaches in the world to share their insights and improve my skills. I hired private coaches to work with me one-on-one, on and off the hill as I pursued perfection. My mind was perpetually open to new ideas, training methods and technical drills. I did not let myself reach a level of satisfaction, because, like in nature, the process of perfection never ends.

In engineering, I brought my "defy-the-odds" attitude from the mountain to the classroom, the lab and the offices of my professors. At Colorado School of Mines, there is no slipping through the cracks with a diploma in hand. Professors do not accept mediocrity. Students drop out. I recall 24/7 weeks of training, lab work, book work, repeat. Often, professors would not accommodate my World Cup schedule, which forced me to decide between pushing classes out to future semesters or giving up. Giving up was not an option. I eventually earned my chemical engineering degree with honors. This unique combination of World Cup skiing and education in the hard sciences has fueled and propelled me into the work I do now. I realize it was and still is all about energy and frequency.

The reason it is important to protect your energy from destructive interference is that you must match the frequency of the reality you want before you can achieve it. This is something I think is often missed or overlooked when people try to utilize the Law of Attraction. The standard definition of the Law of Attraction is the belief that by focusing on positive or negative thoughts, a person brings more positive or negative experiences into their life.[8] The belief is based on the idea that people and their thoughts are both made from "pure energy," and that through the process of "like energy attracting like energy," a person can improve their health, wealth and personal relationships.[9]

I believe there is more to it than that. Yes, the thought or image is where everything in the physical reality originates. However, without the feeling and emotion of having that reality, the thought is just a passing thought that does not materialize. In other words, it is the emotions that carry the frequency of that reality! The very word "e" motion originated from "energy in motion." Thoughts are tied to emotions, but it is possible to have a thought without emotions tied to it or a thought with something blocking the emotions associated with it. In these cases, the thought cannot manifest into material form because you have to match the frequency of the reality you want. You could say that our emotions are a gauge of what frequencies we are emitting.[10]

---

8. S. Whittaker, Secret Attraction (The Montreal Gazette, 2007).
9. Guy Redden, Magic Happens: A New Age Metaphysical Mystery Tour (Journal of Australian Studies, 2009).
10. Gurudas, Flower Essences and Vibrational Healing (Cassandra Press, California, 1989).

Think of two guitars sitting across from one another in the same room. Pluck a string on one guitar, the corresponding string on the second guitar will start to vibrate. Eventually, it will die down, and the original string on the first guitar will start to vibrate again. This is the Law of Attraction, meaning that the first string received back the energy in motion or the frequency that it put out. It is also the Law of Entrainment or Entanglement, where the second string became entrained with the vibration of the first string. You could have the thought of plucking a guitar string and it making a sound, but without actively putting the energy in motion, you would not get the sound back.

As I said earlier, my skating skills supported the transition to skiing. In my

"Everything is energy and that's all there is to it. Match the frequency of the reality you want and you cannot help but get that reality." ~ Albert Einstein

first year of competitive skiing, I somehow won almost every competition I entered. When I discovered that I qualified for Junior Nationals, my first thought was how I would afford the airline ticket. This is where I fortified my view that there are no excuses in life and I would make it happen by pure desire. I immediately began visualizing myself there and feeling the emotion of being there and skiing well. This was my first real experience with matching the frequency of the reality that I wanted.

Throughout my long career in professional moguls, I continued to visualize things into fruition. During a season we would have about 13 World Cup competitions, often in another country every weekend, for 13 weeks straight. We would typically arrive at a venue and have three to five days to familiarize ourselves and practice on the course before the competition. My teammates and competitors from other countries would lap top-to-bottom runs, a run from the top of the course with two airs to the bottom, every training session.

In comparison, I rarely made a top-to-bottom run during my practice on the course. This was not because I was not trying, or I did not want to, but because I could not do it no matter how hard I tried. Either the top air was too tricky to ski into, the bottom air landing was too hard to stick or the turns in the middle were too fast, too steep or too deep. But come competition day, I would always allow myself ample time to sit up at the top of the course, usually in the trees, before my competition run.

I would sit there for at least 30 minutes, often longer, and continuously visualize myself skiing perfectly. I would intensely visualize every detail of my run using all five of my senses, including the "scent of competition" as mentioned earlier. I would see it, hear it, smell it, taste it, but did I always feel it? I was often too scared to even look down the course from the start gate because of the way the top air landing would drop away into the clouds and the moguls were huge, hard and icy. However, almost every time they started the countdown, I pushed out of the gate and skied my best run of the competition week. Believe me when I tell you this was to the amazement of my coaches, my teammates, my competitors and even myself! Out of 150 World Cup starts, I did this 98 percent of the time. I would literally visualize my run into fruition.

What about the two percent of the time this did not happen? In every competition I entered, I always practiced my intense visualization routine, so what went wrong? Looking back now, it is clear to me what was different. I was not able to match the frequency of the reality I wanted. In those few instances where I did not stick my run, I did not feel what it would feel like to ski perfectly. In other words, I was not truly happy before the competition run, so the frequency (energy in motion) vibrating around me was not one of happy feelings.

In these few instances during the World Cup, I clearly remember there being outside negative influences. For example, at the 2010 Olympics, my husband was removed from the training area by the local police, detained and threatened to be deported. The U.S. Ski Team did not provide him with Olympic credentials like they had in the past.

My husband had, at this time, been traveling to all my World Cup competitions for over seven years. He could ski, but not at a competitive level; however, he was very smart and knew what to look for in my skiing which made him one of my best coaches. There had never been an issue with his presence at any of my World Cup competitions, including the 2006 Olympic Games. But when we arrived for the first training session of the 2010 Games, the "officials" would not allow my husband and my coach access to the mountain. My husband was not allowed to stay with me in any of the provided housing nor was he allowed to view my training sessions.

The Olympics had always been such a big deal to me, and we had worked so very hard to get there. At that time in Freestyle Mogul Skiing, only .005 percent of women in the U.S. who wanted to compete in the Olympics actually completed the journey. The process of qualifying was very competitive and mentally and physically exhausting, not to mention the pressure and demands once there.

Our story was picked up by many international stations, but not in the U.S. In an interview with a Japanese station after a training session, I was explaining the situation. I reached in my pocket and pulled out beautiful rose petals Mike had placed there to remind me that he was still supporting me even though he was not there physically at the bottom of the course. Needless to say, this was one of those competitions where I did not stick my best run. I was not able to match the frequency of the reality I wanted without my husband and coach by my side at one of the most prestigious competitions of my career.

Another one of those times was World Championships in 2005 in Italy. I had a rough crash the week before and severely bruised the bones in the knee joint. This may not sound that bad, but considering I had no cartilage left in that knee, my bones would grind bone-on-bone when I skied, making it excruciatingly painful to ski moguls. I managed to put the pain out of my mind and ski well in the qualifier run, but by the time the final run had come I was exhausted and could not maintain the vibration of the reality I wanted.

My third example is from a World Cup competition in Norway where I had bad food poisoning. I had, without realizing it, eaten moldy rye bread the night before the competition. I was very sick the next day, vomiting and hallucinating, but had to compete. Not competing was never an option, so even though I could barely stand I managed to ski well enough in semis to make finals, but I did not ski the way I wanted in the finals. Clearly, I was too sick to match the frequency of being happy and satisfied. However, I was relieved to have survived the day.

I can now see why I was not able to match the frequency of the reality I wanted at certain times in my career and how this affected the outcome. This is not an attempt to blame circumstances; rather, it is a thoughtful reflection on those intense competitive years to try to extrapolate real experiences that are tied to scientific law. In all my years competing on the World Cup circuit, the three instances mentioned above are the only times I fell in my finals run.

When re-examined through the lens of matching the frequency, one can see why. I was intensely visualizing at all competitions, but when I was not happy or feeling good before my run, the results did not meet my expectations. In these instances, I was not able to match the frequency of the reality I wanted.

During my career on World Cup, I learned to always take the day before the competition off. This seemed unusual to people but was a definite contributor to my success. My teammates, coaches and friends would often say things like: "I heard they are changing the bottom air tomorrow, and we think you should come out for just a little bit and check it out;" or "They completely remade the top section of the course."

But I remained diligent in taking time away from the course the day before the competition. Sure, this allowed my physical body time to rest, but even more importantly, it helped me match the frequency of the reality I wanted.

The day off was always spent doing things that made me "feel" good. This usually entailed some form of shopping and looking at pretty, sparkly things. I would not recommend this activity to everybody, but this was one of my favorite things to do. I also always took a quiet walk alone in nature to realign my energy and let go of anything I was holding on to that was not serving me.

Meanwhile, my competitors were out training and often getting frustrated. The day before the competition the course was usually in the worst condition, which caused stress among the athletes training. This stressful environment could spiral into negative thoughts and feelings. This played a significant role in the fact that on competition day, I would often beat competitors who had better technical skills. My frequency was higher because my thoughts and, more importantly, my feelings and emotions were more positive.

Recall Dr. Emoto's and Dr. Tiller's work. They proved through experimentation in the lab that words, or more importantly, thoughts have energy or frequency that alter the molecular structure of water. But, is it really the thought? Or, is it the emotion associated with the thought?

Jim Self, a former advisor and Director of Governmental Operations for the Department of Energy, an international author and speaker on energy, ties thought to emotion by saying that thought is the electric energy and emotion is the magnetic energy. James Clerk Maxwell was an Einstein/Newton-level genius who took a set of known experimental laws (i.e., Faraday's Law and Ampere's Law) and unified them into a symmetric coherent set of equations known as Maxwell's Equations.*

Through Maxwell's Equations we see that light or energy travels in a propagating wave of electric energy that gives rise to magnetic energy

---

*Maxwell's equations are a unification of Faradays Law and Ampere's Law stating that:

1. A flowing electric current (J) gives rise to a Magnetic Field that circles the current
2. A time-changing Electric Flux Density (D) gives rise to a Magnetic Field that circles the D field

Ampere's Law with the contribution of Maxwell nailed down the basis for Electromagnetics as we currently understand it. And so we know that a time-varying D gives rise to an H field, but from Farday's Law we know that a varying H field gives rise to an E field... and so on and so forth and the electromagnetic waves propagate.

that gives rise to electric energy and so on and so forth. If Jim Self is right that thought is electric energy and emotion is the magnetic energy, then one gives rise to the other as emotion and thought move along.

I do not know of any scientific data backing up Self's hypothesis, but let's take a closer look. My view is that thought is potential energy, i.e., energy stored in an object that represents a multitude of potential. Once this potential energy is set in motion by an "e"motion, the potential energy is transformed into an electric current. As stated above, this gives rise to magnetic energy and the two feed each other as the energy propagates along. When you have a passing thought, it is pure potential energy.

However, once the thought connects with emotion or magnetic energy, it starts to move along at a certain frequency. That frequency can increase or decrease depending on the intensity of thought and the feeling it evokes.

Thoughts and emotions gain in intensity and momentum leading up to the competition day where they reach their peak. In other words, as I would get closer and closer to the competition day, my thoughts on how to ski the run perfectly would increase and my emotions tied to the physical performance would intensify. On the day before the competition, this process creates a state of mind crucial to success in the start gate on competition day. If thoughts move down a negative path at a time when they are gaining momentum and intensity, these thoughts can start to snowball in a bad direction fast.

Therefore, the more crucial training to be done the day before the competition is much more mental than physical. Sometimes shifting a train of thought can be very difficult, especially if there is momentum behind the thoughts. If I had a bad day of training I would shift to a completely unrelated thing to focus on, i.e., something I knew would make me feel good like the shopping mentioned earlier. In doing so, I would shift my emotions first, then having more positive thoughts about skiing was much easier to achieve. As a result, tuning into the right frequency of the reality I wanted on competition day became much more attainable.

As you can imagine, skiing well made me very happy, especially at big events like the World Championships, the Olympic Trials or the Olympic Games. The bigger the event, the more I was to gain from skiing well and the more elated I would be. This, of course, made skiing well at big events even harder because I would need to be in an extremely elated state before the competition run to match the frequency of the reality I wanted. I knew this on a subconscious level when I was competing, but not on a conscious level like I so clearly see now.

If I had understood this more thoroughly, I am sure I would have done a few things differently. For example, at the 2006 Olympics I skied well in the finals, but I did not stick my run. I had a bobble or mistake in the middle of the run. This was very uncommon for me but, given the emphasis put on the event, it made matching the frequency of skiing to perfection extremely difficult. This particular "big" event also attracted family members who had never seen me compete live from the bottom of the course. They were not part of my core support team and carried heavy energy that was difficult to elude.

A good athlete becomes great at protecting their energy from others, but this is much more challenging when it comes from your own family because of the longer-term energy ties you have with them.

I think if organizations really understood this principle based on the scientific Law of Attraction, they would do things differently as well. For example, the U.S. Ski Team thinks it is best to select the Olympic Team as close to the Olympic Games as possible. This seems logical as the goal is to take the skiers that are skiing their best at that time. However, this forces athletes to be strictly focused on achieving peak performance at the qualification events leading up to the Games.

If the athlete waits to qualify for the Games before they start having thoughts and emotions of winning an Olympic medal, they have already lost. Everybody competing in the Olympics wants to win a medal and, for the most part, has the ability to do so. Athletes who land on the podium tend to have the fastest, clearest thoughts and most positive, intense emotions. Some countries pick their Olympic Team a year in advance. From the perspective discussed here, this puts those athletes at a significant advantage for winning Olympic medals.

Before my competition runs, I was often in the lodge crying. Crying is a way of processing and releasing lower vibrations or thoughts – that is, thoughts of not being good enough or not being able to ski from the top to the bottom of the course. It is also a clear sign of intense emotions vibrating around clear thought. Through the action of crying, I could readjust my energy to a frequency of the reality that I wanted of skiing perfectly from the top to the bottom of the course.

The competitions where I performed to my greatest potential were undeniably where I was on edge emotionally, meaning my emotions (energy in motion) were high. For example, there was an important World Cup run called Champion in Deer Valley, Utah, where I had to ski well to maintain my spot and be invited to the Olympic trials. In training I came in short on a backflip. In other words, the tips of my skis dug into the snow near the end of the backflip instead of landing underneath my body. With my body in the air horizontal to the ski hill combined with my forward momentum, I tore my hip flexor.

This was the morning of the competition, and not skiing was not an option. On the World Cup circuit, you had to perform every weekend at every World Cup to maintain an adequate ranking, or you would be sent home. I was in a

Deer Valley, Utah
World Cup Win

Park City, UT
Sporting my prized sparkly Bogner jacket

lot of pain and remember crying on the chairlift on Mike's shoulder. Through my sobbing and sniffles, I said to him, "If I stick my airs on this run and really get a piece of each one, can I have that sparkly Bogner jacket (a gorgeous and expensive ski jacket that looks more like it belongs on an ice skater then a skier)?"

On a subconscious level at that time, I was trying to shift my thoughts to what made me happy and excited, so I could match the frequency of the reality I wanted. I also took some extra time at the top of the run to connect with nature – to the trees, the breeze and the warmth of the sun, which always helped realign my energy and intensify positive thoughts. I did stick the run, won the competition and was invited to the Olympic trials.

There are likely many tales and stories about my unusual ways that I am sure circulated around the table at dinner conversations on the World Cup circuit. There was one thing I did that was not easy to hide, and I did that one thing excessively. In Freestyle Skiing you ski with a pole in each hand for balance and timing through the moguls, but I also used them to clear energy. I did one or two or three or…okay, often many, many "pole clicks" (i.e., the tapping of the poles together either in front or behind the body) to clear the energy not serving me so I could have more positive thoughts that would link up with my emotions.

I am sure I drove more than a few people crazy with my excessive pole clicking. To match the frequency of the reality I wanted, I found I had to be free and clear from any negative thoughts, emotions and feelings.

Another thing I did was sleep. This may sound crazy in a high-pressure environment, but I found it crucial. We would typically have our qualifier run first thing in the morning, and if you finished in the top twelve, you would ski a finals run a few hours later. I was a pro at the five-to 10-minute power nap in the lodge or wherever I could find a quiet place. I was not the only one. I noticed this as a practice of several of the highly successful athletes on the World Cup circuit.

From my experience I found sleep to be a great way of resetting to a place of calm, happy, relaxed energy so that my physical body would not get worn out from operating at a high frequency for too long. There is a lot of research that points to the benefit of sleep to realign your energy or vibration. Again, at the time, I only realized these things on a subconscious level. But looking back, this would explain why it was more difficult to perform well at big events. Big events attracted many spectators and media outlets. It was always difficult, if not impossible, to find a quiet spot to sleep at these bigger events.

At the time I thought these were funny little quirks or odd things I did, but now I know through science what I was doing, how I was doing it and why it helped me perform at such a high level for so long. I hope that this insight will help readers as well. If you are planning an event, giving a presentation, competing in anything or taking a big test, and you keep thinking about all of the things that could go wrong, then let me tell you, they will. If your thoughts cause you to feel sad and disappointed, you are likely to experience outcomes in line with those emotions. If you visualize your event going off without a hitch, being happy or filled with delight and you actually feel those emotions while you are visualizing, then that is what you will get!

This is often easier said than done, which is why I recommend coming up with a ritual of visualization, doing things that make you feel good, taking power naps, releasing emotions through crying and always finding ways of connecting with nature. Remember that your feelings or "e" motions are the frequency you put out more so than your thoughts. I always found that in the absence of any blockages, the more focused and intense my thoughts, the more intense my emotions.

According to the Law of Attraction, you get what you put out in the form of vibrations. In this context, if you want happiness, abundance, or joy, act as if you already have it. Do this by not only thinking of it and visualizing it, but also feeling it with intense focus and consistency. In this way, you will be matching the frequency of the reality that you want.

Conservation of Energy

Another essential universal law to consider is the Law of Conservation of Energy. This law states that energy cannot be created or destroyed, but only changed from one form into another. For instance, chemical energy is converted to kinetic energy when a stick of dynamite explodes.

If one adds up all the forms of energy that were released in the explosion, such as the kinetic energy of the pieces, as well as heat and sound, one will get the exact decrease of chemical energy in the combustion of the dynamite.

If you have a negative passing thought that you do not want, it is important to keep in mind that the thought cannot merely be ignored. Rather, the thought needs to be transformed into a more positive one; otherwise the negative thought lingers around obstructing the emotion tied to the positive thoughts you want.

People who have been having negative thoughts for a while have a much more difficult time shifting them. This is why you can not tell someone suffering from depression to just "think more positive." Not only would that seem so painfully unsympathetic, but it will be very difficult for them to do. Why is it hard to "think more positively in these situations?" The energy body can get so tangled and cloudy from negative thought that it can block the emotions tied to positive thoughts.

Wim Hoff, a.k.a. The Ice Man from the Netherlands and world record holder in extreme sports, believes he has found a cure for depression and other chronic diseases. He has developed a technique for controlling one's body temperature during exposure to extreme cold temperatures with specific breathing techniques. He believes that with regular practice of this technique one can have more control over their thoughts and even their physical health.

> Utilizing bioenergetics found in nature we can make "quantum nutrition" for our energy bodies.

I too believe I have found a way to help people "have more positive thoughts" through a cleansing or clearing of the tangles in their energy body. Utilizing bioenergetics found in the botanical world we can make "quantum nutrition" for our energy bodies. Similar to eating a healthy well balanced organic diet to nurture our physical body, we can nurture our energy body with specific combinations of bioenergetics found in nature, quantum nutrition.

Energy = Mass x Conversion

Competing professionally for 16 years in any sport is hard. I believe that mogul skiing is on the extreme end of the difficulty spectrum. One faces a different course every week, a wide range of snow conditions and it is imperative to constantly push the boundaries of what your physical body can do. To say it is hard on your body is an understatement. What prolonged my career from a physical perspective was the recognition that optimal health does not just apply to the physical body.

Over a hundred years ago, Albert Einstein gave us the natural law of:

$$E = mc^2$$

where E = energy, m = mass,
c = the speed of light

Have you ever thought about what this really means? If you multiply your physical mass times the conversion factor of the speed of light squared (i.e., 3.47×10^{10} miles per second) you get your energy body.[11] Our energy bodies are far bigger than our physical bodies. As an athlete, I would work out in the gym about four hours a day in

11. David Bodanis, E=mc^2: A Biography of the World's Most Famous Equation (Berkley Publishing Group, 2000).

the offseason. My physical body was strong, fast and agile. At the time, I thought I was doing all I could to help prepare and sustain my body for the physical demands of the World Cup tour. But, on a subconscious level, I was doing so much more.

Recall all of the obstacles I had to overcome in the first half of my career, e.g., lack of money, six knee surgeries and numerous critics. Looking back, I can plainly see things were not flowing during that phase. I believe this was because my energy body was not in alignment, like when your back may not be in alignment at times. I was still carrying around those thoughts of, "You cannot do this without the support of your family," or "You are not the elite of the elite," or "You will never ski professional moguls again."

In other words, I was still carrying around thoughts and beliefs that conflicted with what I wanted. These thoughts were tangled up in my energy body. Similarly, to the way I had to work out knots in my tired muscles and ice injuries in my physical body, I needed to work out the knots or tangles in my energy body.

To accomplish this on a more sustainable level, I had to go deeper than just doing the things mentioned earlier like shopping and looking at sparkly, pretty things. I had to release old thought patterns and reprogram beliefs that I did not even know were affecting my physical performance. These things were tangled up in my energy body and needed to be released so I could fully accomplish my dreams.

A shift in my ski career happened when I was about halfway through. Recall that it was on the advice of the Ski Team sports psychologist, that I should visualize with all my senses. I then sought to find a "scent of competition." This was when I started working more closely with the bioenergetics of nature. I believe, and continue to confirm through research and science, that this was a huge contributor to my ability to compete professionally for another five years injury-free.

Up until this point, I was doing all of the intense visualization and of-the-moment energy clearing talked about earlier. However, I needed to clear out the thoughts and beliefs in my subconscious that were still holding me back from achieving my goals. The energy I harnessed from the bioenergetics of nature in my "scent of competition" was helping me do this on a subconscious level. Several years later, through my research at Phia Lab, I was able to show through science and on a much more conscious level the incredibly powerful effect that these natural energies have on the energy body and how that ties to the physical body.

Generation after generation, we have had certain beliefs drilled into us. According to Dr. Bruce Lipton, these beliefs stem from childhood or even former lifetimes and reside deep in our subconscious mind. If the belief was that you were not good enough or not deserving, then these beliefs would hold you back from being good enough and deserving without your conscious awareness of it.

The energy contained in the bioenergetics of nature that I harnessed in creating my "scent of competition" was literally shifting my energy.

The energy contained in the biological essences used in my customized scent was literally shifting my energy, untangling my energy body and helping me vibrate at a more clear, unobstructed frequency on a regular basis. Suddenly, it was not taking as much work on competition days to match the frequency of the performance I wanted. As a result, I was on the World Cup podium more frequently. Training sessions were not as hard on my physical body, so I could train for longer hours without getting tired. Naysayers had little to no effect, and money was starting to flow.

Looking back, this all makes perfect sense. Science has proven that our sense of smell, more than any of our other senses that must travel through multiple sensory locations, is directly related to the area of our brains associated with memory and emotion.[12] Have you ever smelled something and immediately been taken back to a place and feeling in your childhood? Scent has a profound ability to affect our emotional state.[13]

Scholarly research is pointing to a relationship between our emotional state and electromagnetic energy that we emit or our energy body.[14] Furthermore, research is revealing a direct correlation between our energy body and

---

12. Valerie Ann Worwood, The Fragrant Mind – Aromatherapy for Personality, Mind, Mood and Emotion (New World Library, 1996).
13. T. Matsumoto, H. Asakura, T. Hayashi, Effects of olfactory stimulation from the fragrance of the Japanese citrus fruit yuzu (J Altern Complement Med., 2014).
14. Kiecolt-Glaser, Graham and Malarkey, Olfactory influences on mood and autonomic, endocrine, and immune function (Psychoneuroendocrinology. 2008).

the health and wellness of our physical bodies.[15] This is not far from what Ayurvedic practitioners have believed for thousands of years.

What better, more effective, more efficient way to shift our emotions, our feelings and, therefore, the frequency we put out than through the lens of smell? Recall that through the Law of Attraction we receive the frequency we put out, so we want that to be positive, in alignment and as clear as possible.

The most ancient texts of Hinduism, Ayurvedic medicine, and Shamanism have some interesting things to say about the breath. One of these sayings is "nourish

The physical body gets nourishment through things we eat and drink, and the energy body gets nourishment through the energies we breathe in.

15. Elizabeth Blackburn, Accelerated telomere shortening in response to life stress (Proceedings of The National Academy of Sciences, 2004).

your body the way you breathe."[16] For the longest time, this has been thought to be some kind of mystery or secret. Could it be that these ancient civilizations were very much aware of the much larger energy body and that it could get nourishment through our sense of smell or the act of breathing? Like the way our physical body gets nourishment through the physical things we eat and drink, our energy body gets nourishment through the energies we breathe in.

Could it be that in our audio and visual world we have tuned out our most important sense, our sense of smell, and in doing so we have forgotten how to truly nourish our bodies? Sheldon Saul Hendler, MD, Ph.D., and researcher said: "Breathing is the key that unlocks the whole catalog of advanced biological function and development. Is it any wonder that it is so central to every aspect of health? Breathing is the first place, not the last, one should look when fatigue, disease, or other evidence of disordered energy presents itself. Breathing is truly the body's most basic communication system."[17]

More common today than ever before, scientists are starting to show that illnesses are appearing first in our energy body before taking

16. Vladimir Megre, The Energy of Life (Self-Published, 2002).
17. Hendler, Sheldon Saul, MD, Ph.D., Oxygen Breakthrough (Reed Business Information, Inc., 1989).

form in our physical body.[18] This indicates that there should be benefits in treating the energy body before any illness has the chance to manifest in the physical form. Since our sense of smell is directly related to the part of the brain associated with our emotions and therefore, our energy body, it would seem that the use of nature's aromatics would be an effective way to nurture the energy body.

There are other ways to cleanse and nurture the energy body such as yoga, meditation, massage and gratitude walks. Out of these, gratitude walks are my favorite. There is something that feels so good about giving thanks and appreciation for all that you have and for where you are in that moment. Although these seemed to be effective in clearing of-the-moment energy issues, I did not find them to be as effective at clearing deep-rooted issues in my subconscious as using the energy harnessed from the bioenergetics of nature.

Of course, there were all of the obvious things like physical strength, agility, eye training and eating right that contributed to my athletic success. But, it was the recognition and nourishing of my energy body that kept me "fit" in mind, body, and

---

18. Susanne Babbel, Ph.D., Somatic Psychology - The Connections Between Emotional Stress, Trauma and Physical Pain (Psychology Today, Sussex Publishers, 2010).

energy. Ultimately, this is what allowed me to have such a long and prosperous skiing career.

You may be super busy taking care of everyone else's needs, or you may be caught up in your job or other day-to-day activities, but it is important to remember to take care of yourself. This covers not just your physical body, but your much larger energy body because it has a direct effect on your physical health and wellness, and whether or not you accomplish your goals. In fact, I think it is more important to have a membership to a spa or wellness facility that helps you work out your energy body than it is to have a membership to a gym to work out the physical body.

Above all else, I encourage you to recognize the benefits of nature's essences. This could translate into more frequent escapes to natural places and bringing nature back into our concrete jungles with the use of biological essences. In nature, we have the opportunity to "nourish ourselves the way we breathe" and take in the beneficial vibrations of the natural world through our sense of smell. Any way you can bring nature into your personal story it can provide you with a formula for success and a well-lived life!

Our research at Phia Lab seeks to contribute to this broader approach to wellness by connecting the dots between the bioenergetics of nature and their ability to nurture our energy body and therefore, our physical well-being.

Chapter 5
Nature's Secrets: A Deeper Dive

After the 2010 Olympics, I retired from skiing. Mentally, I was prepared for more discovery as the last few years of my skiing career had opened my mind. I wanted to take a much deeper dive into my experience with the "scent of competition" and became very interested in understanding aromatherapy from a scientific perspective.

Aromatherapy is an ancient form of medicine that utilizes the biological essences of plants or highly concentrated aromatic plant essential oils. I had experienced firsthand the effect these essences had on me in terms of elevating my performance and sustaining it at a high level. I was on a hunt for answers in a similar way to how I attacked mogul skiing. I wanted to understand the "why" and unleashed the engineer in me to pursue facts and formulas.

Aromatherapy started approximately 6,000 years ago when biological or aromatic plant essences were first used by ancient civilizations to cure illnesses and improve emotions.[19] Despite this history, today aromatherapy is pushed aside by mainstream healthcare professionals in favor of modern medicine due, in part, to a lack of understanding and scientific research.

When I was educating myself on aromatherapy by reading all the books I could find, I quickly became frustrated with the information available. Much of the information was contradictory; there was a lack of scientific data supporting it; and, it seemed there was a limited understanding of how aromatherapy really worked. For these reasons, it is understandable that aromatherapy is not taken seriously as a form of medicine in mainstream health care.

There was a plant biologist and soil scientist by the name of Bruce Tainio who was cited in the majority of aromatherapy books I was reading. He invented the BT3 frequency monitor to study the frequencies of plants. Dr. Gary Young of Young Living knew about Tainio's research, and convinced him that he needed the monitor to study the frequencies of his essential oils. Dr. Young claimed to be able to measure the vibrational frequency of

19. Luca Turin, The Secret of Scent.

fruits and vegetables, healthy bodies and unhealthy ones and the frequency of various essential oils.

This work was published in Dr. Young's books on aromatherapy and cited in many other books on the same topic. I thought this sounded very interesting and more scientific. When I dug deeper, I found out that none of the work was reproducible, and that the device used in the testing had been retired. I also learned that Dr. Tainio never intended the monitor to be used in the way that it was portrayed in various books. Frustrated with this discovery, I started to doubt the validity of explanations in other aromatherapy books as well.

Nevertheless, this did not seem to have hampered aromatherapy's rising acclaim over time. Actually, aromatherapy is gaining more ground and enticing more followers without having the full backing of a clinical study.[20] Imagine the impact aromatherapy could have if there were more clinical studies and research to demonstrate its effect.

What if it were possible to uncover scientific data that could prove a link among attributes of certain known classifications of biological essences? This information could aid in developing a more clinical understanding of how and why aromatherapy works. It could also help expand the application of these

20. Kiecolt-Glaser, Graham and Malarkey, Olfactory influences on mood and autonomic, endocrine, and immune function.

timeless practices from being focused on ailments to the everyday enhancement of life, whether performing on stage on opening night or pushing out of the start gate on competition day.

I began with the advice of Albert Einstein who said, "Look deep into nature, and you will understand everything better." I found that nature holds many secrets across a wide space of less-traveled territory. To unlock some of these secrets, I examined biomimicry, the bioenergetics of nature and sacred geometry through the lens of science and engineering.

Biomimicry

Dr. Bruce Lipton provides a good explanation of biomimicry: "Here in the States, scientists are finally recognizing that Nature is the best teacher. For several centuries, humans have held the belief that we were 'created' to have dominion over Nature. In this regard, scientists are finally focusing their attention on other species and how they have learned to deal with life's stressors. Biomimicry, the science of reverse-engineering the secrets of biology, is revealing new ways to use Nature's technology to avoid the crises that we continuously face."

In this new field of study, scientists are looking at nature and natural systems as a whole and trying to duplicate or replicate nature instead of the more common process of extracting from nature. Like Dr. Lipton, I feel that in our technocratic world, we believe that if something is great, a human must have made it. In the words of Janine Benyus, who founded the field of biomimicry: "I was once looking at a beehive, studying its intricacies and marveling at its construction with a five-year-old child. The child looked at me and asked, "Who made that?" When I replied the bees [did], the child was confused. From his technocratic worldly perspective, he assumed a human had to have constructed it."

For hundreds of years, we have been living in a predominately Newtonian world. Rooted in the perspectives identified with Isaac Newton's name, this is a world where everything is systematic, compartmentalized and can be analyzed in parts. It is a world where we extract from nature and recreate its parts in a laboratory, failing to understand the whole of the natural system. I believe a shift in this perspective has come. People are now looking to nature to understand systems instead of parts as in the field of biomimicry. After all, nature does have a few billion years of progress over humans.

The pictures to the right show examples of biomimicry.

1. Clamshell – It is twice as tough as man-made ceramics, and yet it is produced without using any heat. No heat, just self-assembly out of seawater.
2. Spider silk – The highest performance material known to mankind in terms of its durability, pliability, and weight. According to the Biomimicry Institute, researchers are trying to make suits of armor out of it, and it is being used in the medical field to repair tendons and ligaments.
3. Trees – A marvel in moisture transfer. Trees suck water out of the ground and carry it to the highest point transferring it to a leaf without using a single pump. Engineers are starting to use this natural technology in some buildings as a kind of wallpaper that carries water throughout the building without using any pipes or motorized pumps.
4. Peacock feathers – Create color with shape, bending and twisting light as it reflects as opposed to pigment dyes that we use to create color.

It is exciting to see the traction over the last few decades of people trying to understand nature and natural systems. This new field of study has inspired me to uncover the science of how nature is nurturing. Beginning with my experience of living in a tent on the mountainside, I developed an authentic love and appreciation for nature. It was my go-to source of balance on and off the mogul course, and I feel empowered to show my gratitude by spreading that insight through my ongoing research.

Bioenergetics

The standard definition of bioenergetics is the scientific study of the flow and transformation of energy in and between living organisms and between living organisms and their environment. I believe through harnessing the bioenergetics found in biological essences, as a whole and not the sum of its parts, a very powerful energy medicine can be recognized.

We still have a lot to learn from nature. Let's look at the biological essence of rose for example. It takes 60,000 rose petals to make one ounce of rose essential oil. This can be very costly. So, what has been done? Chemists decided to synthesize rose essence in a laboratory. Large companies acquired this information and started manufacturing "rose oil."

If a chemist is reproducing the biological essence of rose oil in a laboratory, the following will likely take place. The chemist will identify the eight to 10 key molecules she thinks primarily make up the scent of rose, synthesize it in a laboratory and present the product as the rose fragrance oil. This is the standard practice of extracting from nature and recreating its parts in a laboratory that has commonly been used. Some would say these synthetic versions of the biological compounds found in nature are not bad because they are composed of the same molecules that occur in nature. While this is true, there are three important facts that I learned from

It takes 60,000 rose petals to make one ounce of rose essential oil.

studying biomimicry and bioenergetics that are important to point out:

1. There are many more than 10 molecules found in the biological essence of plants.
2. Synthetic compounds lack homeostatic intelligence, i.e., cell intelligence that works toward restoring and maintaining balance within plants.
3. Synthesized compounds contain isomers, which are molecules with the same chemical formula but slightly different arrangements than the molecules they are intended to represent.

There are not just eight to 10 molecules in the biological essence of plants. There are hundreds. This means that the synthetic is missing the vibrational energy from hundreds of molecules and lacking the medicinal effects of the minor chemical constituents. If we look at the human body, for example, phosphorus (P) is present in amounts less than 0.5 percent, but it is vital to survival. There is one P atom on every nucleotide unit in DNA. Without it, we would have no cellular intelligence. Iron (Fe) is another example, as it is present in the body in amounts less than 0.05 percent, and yet, it is vital for our body to carry oxygen in the blood. Fe and P provide evidence on how critical a trace element can be, making the difference between life and death. It is the same with the biological essence of plants created in nature, as their healing powers can be driven by trace components within the essence.* In other words, don't expect to feel a lasting energy benefit from a synthetic fragrance off the counter at the department store.

Another example of the importance of trace components found in plant essences is in the case of myrrh. Myrrh contains many compounds that are individually toxic, and yet, myrrh is one of the safest, mildest and gentlest oils in nature. In ancient times, myrrh was used as a sunscreen despite its high concentration of furanoid compounds, which are known to amplify ultraviolet light. According to Dr. David Stewart Ph.D., D.N.M, who wrote the book *Chemistry of Essential Oils Made Simple,* myrrh was "the oil most commonly used in biblical times, and it is mentioned more than 150 times in the Bible," despite containing traces of acids, like formic acid. Formic acid is the poison that causes the itching and burning from many insect bites.

---

*A trace element is a chemical element whose concentration (or other measure of amount) is very low (a "trace amount").

Dr. Stewart also explains, "myrrh contains 3% xylene, which is listed in the top three toxic compounds to look for in hazardous waste sites. However, myrrh also contains 60-80% terpenes which are compounds well known for their mitigating, coordinating and quenching qualities."

Thus, myrrh is an excellent example of how chemical compounds that are dangerous alone can be safe and beneficial in the company of other compounds that mitigate their harsher properties.[22] In other words, myrrh is completely safe and beneficial when utilized as nature made it, but harmful when common practice extracts it into its parts.

Have you ever wondered why the biological essences extracted from plants are often labeled as "essential" oils? Essential oils are *essential* to the life of the plant. The oils travel through the plant with homeostasis intelligence within the cells.[23] This homeostasis is cell intelligence that knows when the plant may need more or less sun, more or less water or if it has a cut that needs to be repaired. This intelligence also offers protection from viruses, bacteria, parasites, and fungi.

22,23. D. Stewart, Chemistry of Essential Oils Made Simple.

Ultimately, homeostasis works towards maintaining balance or equilibrium within the plant, where every vital biological process within the living organism is functioning as it should.[24] According to Dr. Stewart, this homeostasis maintains balance, first in plants that create the oils and then in humans who apply them.

This means biological essences created in nature can work in different directions depending on the needs of the plant or person. For example, oregano oil (*Origanum vulgare*) will kill hostile microbes while nurturing those that are friendly. Myrtle oil (*Myrtus communis*) is an adaptogen that can stimulate an increase or decrease in thyroid activity depending on a person's condition. Synthesized compounds lack homeostasis intelligence and thus are incapable of such intelligent discrimination and act only in pre-programmed directions, whether beneficial or not.

Antibiotics and some pharmaceuticals indiscriminately kill all bacteria in our bodies, including the necessary good bacteria. They also cause our bodies to become more acidic, which can stimulate the growth of fungi and

---

24. D. Stewart, Chemistry of Essential Oils Made Simple.

other organisms in the body – thus trading one form of sickness for another.

The bioenergetics found in the botanical world has none of these negative side effects. It has been claimed that biological essences found in nature have the intelligence to distinguish good bacteria from bad and attack the bad while nourishing the good.[25] Essential or biological oils found in nature work for people in much the same way they work in plants, thus helping to maintain health and balance. Synthetic versions lack homeostasis intelligence and thus cannot deliver health and balance benefits.

In the process of synthesizing molecules, isomers are a common by-product. Isomers are molecules with the same number of atoms of each element but a different configuration of the atoms, Figure 2. In other words, isomers are compounds with the same molecular formula as the molecule they are intended to represent but different chemical structures. This is significant because the isomers can have entirely different properties that can be very toxic. Natural oil of wintergreen is 90 to 98 percent methyl salicylate (C8H8O3), which is completely safe when produced by a plant but

25. D. Stewart, Chemistry of Essential Oils Made Simple.

Figure 2

Example of Three Isomers

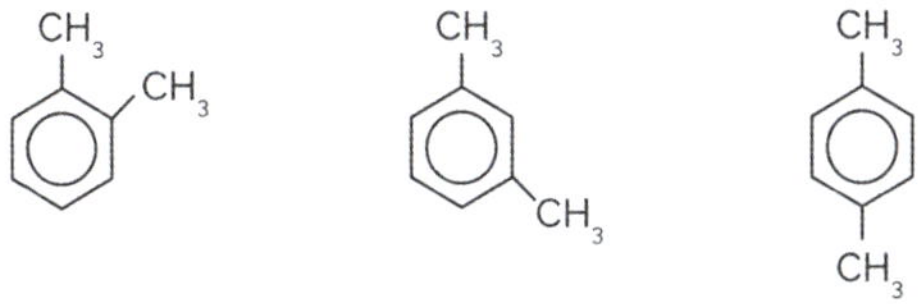

ortho - structure
1,2 - dimethylbenzene
or
o-xylene

meta - structure
1,3 - dimethylbenzene
or
m-xylene

para - structure
1,4 - dimethylbenzene
or
p-xylene

toxic when produced in a laboratory. This is because synthetic oil compounds can have the same chemical formula but different structural formulas and, thus, different effects on our bodies.[26] Methyl salicylate (C8H8O3) has 25 known isomers, each with different properties. When plants produce methyl salicylate, they make a specific single pure isomer, only one out of the 25. When scientists in a lab try to make a single molecule, they end up with, at best, a mixture of two or more isomers.[27] These unwanted isomers can be toxic and harmful to the body, making it the most crucial reason synthesized versions of the biological essences can be bad for us.

26, 27. K.-H. Kubeczka, and V. Fofmacek, Essential Oils: Analysis by Capillary Gas Chromatography and Carbon 13-NMR Spectroscopy (J.P. Hornak, 2009).

Over the past 300 years, we have tended to think that we can produce everything better, cheaper and more efficiently than mother nature. Aspirin is a good example. Aspirin was discovered from Native Americans chewing on white willow bark when they had a headache. Scientists extracted, isolated and evaluated the molecules of white willow bark to determine the molecules primarily responsible for dissipating the pain of a headache. Bayer AG packaged it in a pill and called it Aspirin.

This is how we as a society have done many things over the past centuries. Scientists take the example from nature, separate it from its whole, then engineer it in laboratories. This is done because the recreated molecule can be patented and sold. Unfortunately, we have forgotten that our pharmaceutical medicine originated in nature, and nature's medicine could be used with no toxic side effects.

The harmful effects of synthesized drugs are attracting more attention and review. According to Dr. Bruce Lipton, 300,000 people die every year from prescription drugs. These findings have caused a shift in people's perception. This shift has inspired people to reconnect with the source of all medicine, nature. In doing so, we are unlocking more of nature's secrets.

While aromatherapy has been a form of medicine for thousands of years, I found through my journey into the bioenergetics of plants that aromatherapy fails to get the recognition it deserves for three specific reasons:

1. Aromatherapy tries to explain its effects by referring to the molecular structure of specific active components.
2. Aromatherapy is not taking into account the energy in biological essences and the specific energy of the individuals interacting with the essences.
3. An aromatherapist combines essences in a single blend without regard to the energy of the specific essences.

The current studies of aromatherapy try to explain the physiological effects of essences by referring to the molecular structure of active ingredients or tangible material. This direction is very mechanical and reductionist in theory. It fails to realize and appreciate the effect of the biological oil as a whole, which was created by naturally-occurring botanicals, the true scientists of nature. Recall that myrrh oil is very healing as a whole but can be very toxic when analyzed in parts.

Aromatherapy fails to recognize the connection between the energy content in the biological essence and the energy of the individual. For example, an aromatherapist might prescribe lavender oil to someone having difficulty sleeping. While this may work for many, it will not work for all. We have found through our research that by approaching aromatherapy from an energy standpoint, one can understand that we all have a unique "signature" frequency. And, through science, we can understand the difference between constructive and destructive interference.

From this perspective, lavender essential oil will work well to help people whose signature frequency is constructive to the frequency of such oil. However, lavender will have little effect, or possibly an adverse effect, on someone whose signature frequency is destructive to the frequency of lavender.

If a person needs extra energy for a big presentation or event an aromatherapist might recommend some citrus essential oil. This same person might like the scent of jasmine so the aromatherapist may add the biological essence of jasmine to the blend. From our research at Phia Lab, this may be counter-productive if the citrus essence and the jasmine interfere with each other in a destructive way.

With his ground-breaking research, Dr. William Tiller describes a coupler device, as a device created to help transform energy into something material. He has drawings and specifications on perfecting and building a coupler device mechanically. This coupler is a way of bridging the gap between the physical or tangible and the energy surrounding us.[28] This is a way to get to a more coherent mind, body and soul space. But I believe this coupler device has already been made and perfected many times over for billions of years. It is one of nature's many secrets. It is in the bioenergetics of the botanical world.

The physical matter from plants carries enough energy to act as the coupler or the bridge between the tangible and the unseen energies all around us. If we can learn to harness these energies constructively, these energies can align the energy body or increase order in the energy body as well as the physical body.

---

28. Tiller, Ph.D., Dibble, E., Jr., Ph.D., White Paper I - A Brief Introduction to Intention-Host Device Research,0; W.Tiller, Ph.D., White Paper XXIV - An Introduction to Psychoenergetic Science & Information Medicine. (Tiller Institute, 2012)

As a result of this greater coherence, the physical body reaches a heightened state where optimal physical performance can be achieved. At least this seemed to be my experience during World Cup competition for many years. It is this belief and questions from above that led to my in-depth research and findings, accumulated and summarized in the research portion of this book.

Sacred Geometry

Recall that five-year-old girl who wanted to be a chemical engineer. That same girl was always obsessed with numbers. In my earlier skating years, I always did things in three, five and eight combinations. Later, during my World Cup years, everything I did revolved around certain combinations of numbers. About halfway through my World Cup career when I was developing my "scent of competition," I naturally evoked these same number combinations.

As part of my studies post-skiing, I discovered that nature's secrets stretch beyond the botanical world and into the mathematical divinity with which the natural world was constructed. This mathematical divinity is often referred to as sacred geometry or simply the number Phi, (1.618…).

Phi is a number known to mathematicians and engineers as the world's most perfect number. Things in nature grow in Phi proportions. You can see Phi in nature in the way things grow like pine cones, a nautica shell, sunflowers, a tree, and even the human body. In fact, our DNA is a perfect Phi spiral. Phi has been proven to be pleasing to our sense of sight, Da Vinci used Phi in his artwork. Dr. Stephen Marquardt, a highly regarded plastic surgeon, measured hundreds of parameters of the face and applied the golden ratio to establish the outlines of the ideal face.

The "phi mask," as it is known, became a clinical assessment tool to determine attractiveness in human facial features. Phi has also been proven to be pleasing to our sense of hearing, Mozart used Phi in his music. The Stradivarius violin was constructed using precise Phi proportions. Basically, Phi is the mathematical root of all other sequences. It is in every living thing in one way or another, even the galaxy – found throughout microcosms to macrocosms. All life strives towards Phi or perfection. The number combinations I was obsessed with were tied to Phi in that the ratio of the combinations isolated around the number Phi.

Phi the divine proportion

This mysterious form of numeric perfection shapes our world and fulfills our senses. From Egypt's inspiring pyramids to Da Vinci's majestic masterpieces to the soothing sound of the violin, Phi empowers creation.

Phi (Φ), the world's most perfect number (1.618), is evident in beauty, nature, art, music, architecture, and even the human anatomy.

Unlocking some of nature's secrets through this in-depth look into the new study of biomimicry, bioenergetics and sacred geometry have continued to inspire and elevate our research at Phia Lab and the energy products we create today.

The two main goals I had when I was younger seemed unrelated at the time but have culminated perfectly into what I do today. My years on the World Cup circuit taught me to be tuned in and turned on to the unseen energies around me and how these energies could enhance or hamper my physical performance. My education in hard science taught me an in-depth knowledge of tangible, physical matter on a microscopic and macroscopic scale. I also learned to question things and not be afraid to explore the unknown.

Just to say the words "energy products" may seem like such a contradiction to some as our society has learned to separate the nonphysical and the physical over the course of hundreds of years. Given my background, I experienced the effect of the energy around me. I have taken a hard look into nature and tangible matter through the lens of science and engineering to understand how the two are connected, resulting in energy products that enhance well-being.

Chapter 6
Bridging the Gap

It is my hope that we as a global society will continue to unlock nature's secrets and, in doing so, bridge the gap between the two divided attitudes that have prevailed for so long: the approach that is only mechanical, material and compartmentalized as opposed to that which accounts for energy, frequency and wholeness.

After taking the advice of Einstein and diving deep into nature to try to understand things better, I developed a more thorough appreciation for how and why tactics I adopted during my years on the World Cup tour helped me perform well. However, I was also eager to prove things on a scientific level. As I tried to conjure up support for my research in the traditional academic structure, I found that there was little to no interest. The divergence in thinking was starkly evident.

Non-Traditional Research

The battle to gain support for my research idea was similar to my experience skiing when I was forced to change my mindset from "spinning" to "flipping" in the mogul field. I had learned to spin in skating at a very early age. It translated well to mogul skiing in the form of 360s and eventually 720s. My signature jump through my career was what became known as the "Denver Bronco." The Bronco was a 360-degree spin with a spread-eagle facing uphill at the 180-degree mark. For many years, I would throw a 360 iron-cross off the top air (i.e., a 360-degree turn in the air with your ski tips pointed down and in an "X" position) and a bronco off the bottom.

Then, someone came up with a new idea. To advance jumping sophistication and overall entertainment value of the sport, skiers would no longer be allowed to use two vertical 360-degree spin jumps in a single run. I still have a slight grudge against my friend and Olympic gold medalist Jonny Moseley for instigating this rule with his amazing acrobatic abilities on the slopes. As a result, I faced the need to learn how to flip in the moguls.

As a skater, I did not flip, although it sounded easier to do on ice than in the bumps. As you may expect, for someone in their late 20s, fear initially dominated my thoughts as my mind resisted…the takeoff, the landing, the line of sight, position of skis and boots, hands and head, and more. And, it was difficult to avoid the thoughts of injury. After months of trampoline work, water ramping and hiking the jump portion of the course to practice the trick and landing on snow, my mind came around to accept the new reality. There was no other way forward.

That is exactly how I persevered with my research idea. I had the opportunity to go to a well-respected but traditional and conservative engineering school with a heavy focus on the oil and gas industry. The percentage of women at the Colorado School of Mines at the time I started was around 15 percent. As a blond mogul skier with a love of flowers, feathers and rainbows talking about the *energy* in the biological essences of plants, I ran into a few naysayers. It is easy to imagine how this would not be taken seriously. Fortunately, I managed to grab the attention of a few key professors who were willing to step outside the box and embrace some new ideas.

I am thankful to these yaysayers who could put the standard guidelines of academic research projects aside to push the envelope and come to new understandings that are capable of equipping students with the tools to solve 21st-century problems. However, I was definitely surprised at the challenge this presented to most. It seems that the divergence of thinking stems back to the time when we decided to separate the nonphysical and science. In doing so, we failed to recognize the unseen energies around us from a scientific perspective.

The Return of the Alchemist

Hundreds of years ago, an alchemist was someone who tried to refine something to its purest and simplest form to understand the interconnection of the macrocosm and microcosm. Said another way, an alchemist is someone who is focused on liberating something from its fixed physical properties and vice versa. Some alchemists of old attracted criticism with their focus on turning common elements into gold or silver. Other alchemists were masters of bridging the gap between the nonphysical and the physical.

Through the ages, alchemists have concerned themselves with trying to discover a universal cure for diseases and a means of improving and prolonging life by studying the nonphysical and the physical. With the separation of church (nonphysical) and science (physical) in the 17th century, the alchemist, who appeared to be studying both, could no longer continue to do so.

The alchemist had to choose to be a scientist or a priest or be killed. I think we are in a new age where the alchemists of the 21st-century are re-emerging to bridge the gap that has existed for hundreds of years between science and nonphysical phenomena or things one can see and energies that are unseen.

When you study matter and bioenergetics, you cannot help but cross the line that has separated the two areas of study for so long. Nikola Tesla, the famous physicist and engineer of the early 1900's, who discovered radio waves and is primarily responsible for the technology that produced cell phones and the internet, said "the day science begins to study nonphysical phenomena, it will make more progress in one decade than in all the previous centuries of its existence."

As a society, we are doing better, but there is much more work to be done. I support the mindset that we need to face the facts of new physics or quantum physics and rebuild the pyramidal structure that was once based on Newtonian or very tangible, systematical and compartmentalized ideas.

Dr. Bruce Lipton describes it as follows: At the base of the pyramid there is mathematics, then physics sits on that base. The next level is chemistry, then biology, and finally, psychology (PSI). If the physics changes, everything above it that was built on its principles will also have to change. So far, we have failed to incorporate this new physics into our educational system.

Therefore, the Pyramid of Science, Figure 3, remains fractured. This explains many contradictory views on medicine, technology, and even society. For example, there exist contradictory views on medicine in the U.S. between standard medical doctors (i.e., education rooted in 17th-century science or Newtonian principles) and alternative medicine practitioners, who recognize matter and the unseen energies (i.e., alchemical principles).

In new physics, physicists don't just talk about particles; they talk about fields and waves, i.e., energy. We are made of waves and frequencies and

Figure 3
Pyramid of Science

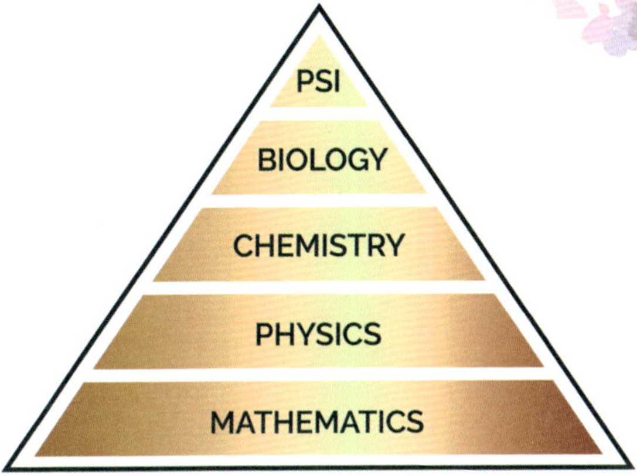

are connected to all things because waves cannot be separated. There appears to be an internal interconnection between ALL elements. Einstein took it a step further and said, "all matter is just light (or energy) slowed down enough to be perceived by our senses."

Since everything is energy on a microscopic level, there is a connection between all things. This is described in physics as the entanglement theory or entrainment that was discussed earlier. Or, according to Einstein, it is "spooky forces that act from a distance."

Is mathematics wrong as well? Now, that may sound a little crazy. One plus one equals two. Everybody knows that. If you pick up one stick and then another, you have two sticks. But if mathematics is looked at through the lens of frequency and waves then it is moving or living from this perspective and one plus one can equal three. Think about it like this, there is a mother, a father and their son. In this case, one plus one equals three. We have about 50 trillion cells in our body that are constantly dying and regenerating. Try telling two cells in your body that one plus one equals two. In this case one plus one equals thousands.

I feel our education system is lacking in teaching both sets of principles to our children. In my public school education, no one talked about energy and the effect it can have on us or that everything is in motion. I learned about energies from nature and my experiences on the World Cup circuit, but imagine the potential we would be giving our children if we taught this in school.

We need more professors in academic circles to start stepping outside the box and embracing new areas of research. These research projects may not be widely accepted by peers and colleagues but are necessary for grasping a deeper understanding of new physics and its broad-reaching implications.

Dr. Elizabeth Blackburn is a Nobel Prize-winning scientist who was criticized for stepping outside the box to do research on how meditation, the practice of mindfulness or the liberating of thought and emotion, affects our physiology. Special gratitude goes to her and many more like her for paving the way for future research in this nontraditional area of study that seeks to understand the seen as well as the unseen. They are preparing the path for the return of the alchemist, someone who recognizes the unseen energy and the visible matter and works with both.

Long ago we forced our scientists to draw conclusions about our physical reality based on what we could see. The visible light spectrum (what we see) is less than 1% of what is all around us. It is not just less than 1%, it is about 0.0035%! And yet we were able to make some gains in physics like trajectories, levers, pulleys, mechanical systems. We also made gains in medicine, synthesizing pharmaceutical drugs and developing early medical procedures.

In the 1800's we started to go quantum with our technology, meaning we started studying things we could not see. At least it was becoming more widely accepted and more readily known that scientists were studying the "unseen." This research produced electricity and later radio waves, microwaves, digital communications and the Internet. Around the same time we started to go quantum with some of our medical procedures developing x-rays, MRIs, ultrasound devices and others. But my question is when did we start to go quantum with our medicine?

As a chemical engineer, I study the bioenergetics of botanicals or the energy contained in biological essences of nature. I hope that this will help bridge the gap between energy and matter, bringing a new wave of research and understanding into the field of energy medicine and quantum nutrition. I believe that this will have wide applications and implications for the use of botanicals in helping everyone maintain the best of health and well-being.

Chapter 7
Research – New Science

It doesn't matter whether you are on the Champion course at Deer Valley or on Riflesight at my home resort of Winter Park, you must lean your body (and mind) forward. Sure, you can make it down the mountain in the backseat, but no one will notice you. Each bump will be difficult to absorb. This will translate into less balance and less confidence. At the bottom, you will be exhausted from the fight and have nothing to show for it.

If you want to get the attention from the six-pack of skiers sitting on the chairlift above, consider gravity a friend and go all-in. Lean forward and allow your momentum to carry you over each bump, shifting your weight from side to side to absorb obstacles with strength and resilience. Reach with your pole to the backside of each bump to feel for resistance before your skis, working in tandem, carve on edge effortlessly through the turn. When you slide to a stop at the bottom, soak up the rush and then glide through the trees over to the next set of bumps to conquer.

At Phia Lab, we have applied this all-in lean forward attitude to filling what I view as a scientific void. We have been diligently contributing to peer-reviewed research into the energy content of biological essences. I owe it to nature in return for the benefits I received, whether it be from my quiet place at the top of the course or from my use of biological essences to evoke the zone. I want to be able to explain how these benefits are delivered to those who are open to receiving them. I took my question to the lab. Hang in there with me as this gets a bit technical.

Biological essences or highly concentrated aromatic plant essences that are used in aromatherapy are known to have varying medicinal properties. It has been claimed that these essences facilitate physiological properties such as relieving pain, combating bacteria, fighting viral infections, relieving indigestion and improving circulation. Such essences have also been connected to social and emotional properties such as easing nervous tension, grounding, increasing concentration, calming and balancing, enhancing creativity and boosting confidence.[29]

29. Kiecolt-Glaser, Graham, and Malarkey, Olfactory influences on mood and autonomic, endocrine, and immune function.

My research efforts into the workings of biological essences lead me to the following conclusions:

1. Biological essences have a unique, distinguishable and quantifiable electromagnetic energy.
2. The distinct energy of the essences seems to be tied to emotional properties.
3. We were able to group essences into six different categories or Energy Waves based on their energy fingerprint.
4. A code of major and minor chemical constituents can represent each energy category of essences.
5. The energy categories of essences can be matched using a "best fit" model to one's signature frequency.
6. There is a positive effect on the energy body of a person interacting with a category of essences with similar frequencies that are constructive to their signature frequency.

The strong focus on energy throughout this research is contrary to the academic exposure of traditional doctors today and even aromatherapists. Through an examination of the energy content of biological essences, we have developed a more complete understanding of how and why these essences have been known to demonstrate a positive effect on our health and wellness.

As part of this work, a new set of concepts and terms were created to better convey findings. One of those new concepts is the Energy Wave. Think of an Energy Wave as a category of approximately 20 biological essences from the botanical world with a similar energy measurement or fingerprint. Essences from the same Energy Wave are combined using precise Phi proportions. These compilations have an amplified frequency that we can align with a person's signature frequency in a constructive way as medicine or nutrients for one's energy body. Similar to the way food, a specific vitamin or supplement nourishes the physical body, the correct Energy Wave can nourish the energy body providing a type of quantum nutrition.

Electromagnetic Energy of Biological Essences

There are thousands of molecules in therapeutic grade plant essences that make up hundreds of chemical compounds, all of which contribute to the overall energy of the essence.[30] Orange essence (Citrus sinensis), for example, contains 34 alcohols, 30 esters, 20 aldehydes, 14 ketones, 10 carboxylic acids and 36 varieties of terpenes, including mono-, sesqui-, di- and tetraterpenes.[31] However, this is not a complete list because these essences are too complex to analyze all their constituents completely. No biological essence has ever been completely analyzed to reveal every constituent.[32]

Therefore, it is virtually impossible to generalize potential medicinal attributes based only on the properties of the known chemical constituents. Trying to predict the essence's properties based on its individual components is a bit like trying to describe the taste of a cookie by the taste of a raw egg or flour. In other words, the biological essence as a whole is more effective than its principal constituents.

Typically, the practice has been to deconstruct the essence into its known major components (alcoholic, phenolic, terpenic or ketonic compounds) and analyze these parts separately

---

30. D. Stewart, Chemistry of Essential Oils Made Simple.
31. Mailhebiau, Philippe, Essential Oil Constituents (The International Journal of Aromatherapy, 1995), Vol 7. No1.
32. D. Stewart, Chemistry of Essential Oils Made Simple.

to find the active ingredient contributing to the medicinal effect. Recall that our bodies only require 0.05 percent of iron to function effectively. It is a very minor chemical constituent, but without it we would die. Similarly, the minor constituents in these essences cannot be ignored in the analysis as has traditionally been the practice.

The relation between composition and activity leads us to suggest that the biological activity of the essences from the plants studied may be attributable to a combination of the major components and the minor components present in these essences. It is possible that they act together synergistically to contribute to the totality of the medicinal properties of the essence. Therefore, the traditional practice of analyzing solely the major components, apart from the whole or a combination of components, is not indicative of its medicinal properties.

From this perspective I could see the need to fingerprint an essential oil as a whole in order to explain how and why they can be beneficial to our physical and emotional health. Unfortunately, there is no analytical technique today that is capable of identifying every chemical constituent in these essences, therefore, allowing for a precise calculation of the overall energy which would be the preferred method. Several methods were considered for evaluating or fingerprinting these essences:

1. Measure thermal conductivity through the essential oil by dividing the magnitude of the electric current density by the magnitude of the electric field (i.e., the inverse of the resistance).
2. Estimate the surface energy of a drop of essential oil by measuring the contact angle.
3. Corona discharge – The placing of the essential oil in an electromagnetic field to extract electrons, and to a certain degree, photons, from the surface of the object. The emitted particles accelerate in the electromagnetic field, generating an electronic avalanche. The discharge gives off a glow due to the excitement of molecules in the surrounding gas. This glow is measured in a variety of ways.[33]
4. Measure spontaneous emission of electrons using a Photomultiplier, an extremely sensitive detector of light in the ultraviolet, visible, and near-infrared

33. David G. Boyers, and William A. Tiller, Corona discharge photography (Journal of Applied Physics, 1973).

ranges of the electromagnetic spectrum.

5. Gas chromatography-mass spectrometry – Gas chromatography works by taking a mixture of gases and moving it through liquid, solid or gas states to separate the gases. This process separates the mixture into its different components, which are then processed through a device such as a mass spectrometer, which describes the individual molecules.[34] This method is good for determining what substances make up a compilation as well as how much of that substance is present. Therefore, this method would be good for analyzing individual components of an essential oil, but not good for outlining the essential oil as a whole.[35]

After careful evaluation of various methods, I settled on examining the corona discharge of the essences. Corona discharge is an electrical discharge brought on by the ionization of a fluid surrounding a conductor that is electrically charged. I constructed a device to do just that. A gas chromatography-mass spectrometry system

---

34. Daniel C. Harris, Gas Chromatography: Quantitative chemical analysis (W. H. Freeman and Company, 1999).
35. Phil Price, Standard definitions of terms relating to mass spectrometry (Journal of the American Society for Mass Spectrometry, 1991)

was also used to gain further insight into the individual components of the essential oils.[36]

The results were surprising. These essences have a variable, distinguishable and quantifiable electromagnetic energy. This was contradictory to basic physical chemistry in that these essences are largely composed of five elements: carbon, hydrogen, oxygen and trace amounts of nitrogen and sulfur. The molecular weights in atomic mass units (amu) of most biological essence compounds range from 136 amu (for monoterpenes) to 240 amu (for sesquiterpene alcohols) and are non-greasy, low in viscosity, quite volatile, and quite aromatic.[37] Therefore, these essences should emit similar electromagnetic energy. But, we discovered a distinctive colorful glow that was unique to each essence, Figure 4. That's right. Each essence has its own signature frequency just like we do!

We tested hundreds of essences and were able to classify them into six distinct categories or Energy Waves according to their energy or frequency. For example, Wave 1 has a wavelength range from 695 nm – 621 nm (nanometers), a frequency range from 4.31E+14 to 4.83E+14 and electromagnetic energy of roughly 182 kJ/mole (kilojoules/mole), as outlined in Table 1.

36. Daniel C. Harris, Gas Chromatography: Quantitative Chemical Analysis.
37. Stewart, Chemistry of Essential Oils Made Simple.

Figure 4
Sample Color Glow

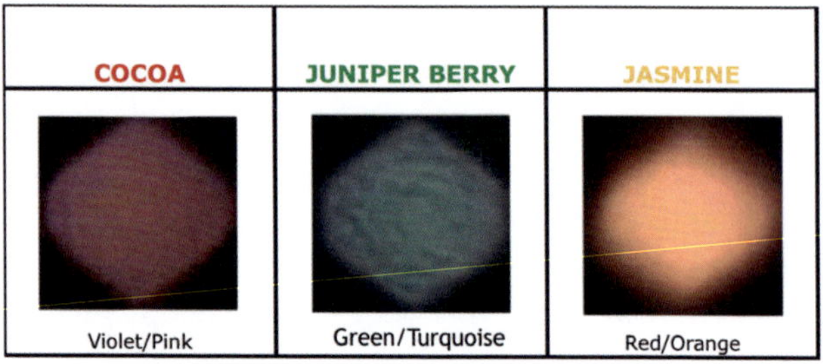

Table 1
Six Energy Waves

	Wave 1		Wave 2		Wave 3		Wave 4		Wave 5		Wave 6	
Visible Color	Brown	Red	Orange	Peach	Gold	Yellow	Green	Turquoise	Blue	Indigo	Violet	Pink
Wave Length [nm]	695	621	620	571	570	521	520	481	480	435	434	410
Frequency [Hz] 10^14	4.31	4.83	4.84	5.25	5.26	5.75	5.77	6.23	6.25	6.89	6.9	7.31
Energy [J/mol] 10^05	1.72	1.93	1.93	2.09	2.10	2.30	2.30	2.49	2.49	2.75	2.76	2.92
Avg. Energy [kJ/mol]	182		201		220		239		262		284	

Emotional Link

What does that mean? I needed more data to understand the results. I started looking for commonalities among the Energy Waves. What this boiled down to was looking at the major chemical constituents, physical properties and emotional properties of each essence evaluated. Oddly enough, the only commonality we found was in emotional properties, pointing to a correlation between these essences that fall into a certain Energy Wave and the known emotional effects that the biological essence could have on a person.

This, again, was contradictory to physical chemistry which would, by definition, point to the major chemical constituents as the commonality in each Energy Wave. I want to point out that not every essence within an Energy Wave had the same emotional qualities, but this was certainly the trend.

I am an engineer at my core, which means I deal with trends and "best-fit practices" as opposed to an exact science. For example, I pulled the emotional qualities of essential oils from hundreds of years of storytelling. These time-tested stories combined with a small amount of clinical studies suggest that lavender helps us relax and that rose boosts confidence. Although over time a significant data pool has been formed, more exact science in the area of how these essences affect our emotions is needed.

Also, the emotional properties of the essences were found from various sources. Often, these sources did not cite scientific studies that led to the stated correlation between the emotional properties and the essence. Information on emotional properties from one source did not always coincide with such information from another source. For these reasons, there is a need for more research that examines the interaction between these essences and their effect on the emotions of the person using them to draw solid conclusions.

However, the underlying fact that these essences have a unique electromagnetic energy that allowed us to classify them and that the groupings tied, albeit somewhat loosely, to the emotional properties of the essence was a compelling finding.

It is interesting that the electromagnetic energy of these essences seems to be connected to the known emotional effect that these essences have on humans. In 1966, Cleve Backster started researching how plants respond to human emotion by hooking his plants up to a lie detector device. He was able to detect a response in the plants right when someone entered the room that varied depending on whether that person had intentions to inflict harm or kindness on the plants. He was inspired to do so after reading about the amazing experiments in the energy fields around plants, trees, humans and cells by Yale Medical School Professor Harold Saxton Burr in the 1930s. Later Marcel Vogel, a research chemist for IBM, confirmed Backster's research by using psycho-analyses, which picked up and amplified reactions from plants.[38]

For thousands of years, essences from plants, flowers, and herbs have been known to influence our emotions, but in what way exactly? Backster's research showed plants can read our emotions. My research indicates that the varying emotional properties of these essences are tied to the essences' electromagnetic energy. Our more recent research seems to indicate that the electromagnetic energy of these essences carries vibrations that can interact with our electromagnetic energy to align or balance it.

---

38. Peter Tompkins, The Secret Life of Plants (Harper & Row, Publishers, Inc., 1973).

Code of Major and Minor Constituents

Since I did not see a commonality in the known major chemical constituents like expected, I decided to expand the scope and look at a representation of all the chemical compounds in each essence to see if there was a particular grouping of constituents that were unique to each Energy Wave. All the essences evaluated were then sent through a gas chromatography-mass spectrometry (GCMS) system. This device shows a peak for each chemical compound in the essence. The height of the peak indicates the concentration of the compound.

There are hundreds of compounds in one essence. For this reason, it would be extremely difficult to identify each compound. However, we were able to overlay the charts to determine a code of six to sixteen compounds unique to each Energy Wave. Figure 5 shows a chart of the major and minor chemical compounds of ylang-ylang, myrrh, helichrysum, lavender, ginger, and lemongrass. These essences all fall into the Energy Wave that is centered around 262 kJ/mol. Given the magnitude of compounds in these essences the printout used in Figure 5 is difficult

Figure 5

Major and Minor Chemical Compounds for Essences in the 262 kJ/mol Energy Wave

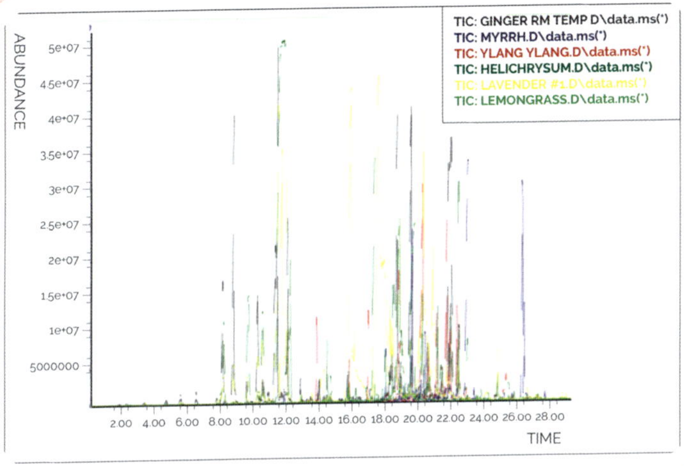

Figure 6

Simplified Code of Chemical Compounds for Essences in the 262 kJ/mol Energy Wave

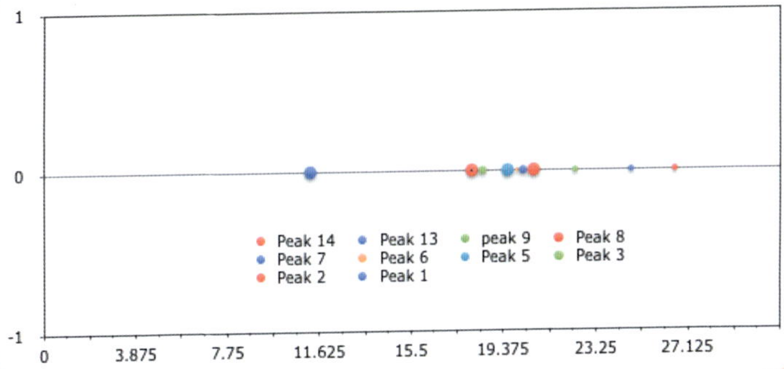

to read in this context. You will find a simplified code of compounds in Figure 6. In this figure each dot represents a consistent peak across all the essences in the grouping. The size of the dot indicates the concentration of that compound.

We concluded that:

1. Major and minor chemical constituent groupings determine the energy measurement for each Energy Wave.
2. These same groupings of constituents could also be the source of emotional commonalities across essences that were found.

Although we could not identify all the chemical compounds in each code, we could potentially use the codes as a way of predicting energy and associated emotional effect of other essences.

Best Fit Model

In our society, we often think more is always better. From this perspective, I usually get the question, when considering the six Energy Waves, "Is the highest category at about 284 kJ/mol the best category for me?" It is important to point out that the energy calculated in this research is the electromagnetic energy of these essences, not the overall energy of the essence itself. To determine the overall energy, we would have to measure the chemical, nuclear, mechanical, gravitational, electromagnetic, thermal and electrical energy of each constituent. This would be very improbable since we cannot identify all the constituents.

We do know that the biological essences from nature vibrate at a clear, concise and higher frequency compared to that of humans.[39] Scientists have found that botanicals have a very similar chemical makeup to humans, but they are much more coherent. For this reason, plants emit 10,000 photons (energy) per second compared to just eight to ten photons per second emitted by humans.[40] Botanicals are very coherent/ordered in their molecular structure whereas humans' cellular structure is very incoherent/chaotic. As a result, a photomultiplier can capture the energy put off by plants, but not that of humans.

The question then becomes which higher frequency category or Energy Wave of essences will resonate best with one's own signature frequency. Our research indicates that in addition

---

39. Gurudas, Flower Essences and Vibrational Healing (Cassandra Press, California, 1989).

40. Peter Tompkins, The Secret Life of Plants (Harper & Row, Publishers, Inc., 1973).

to the many varying frequencies vibrating around us that are constantly changing depending on mood, activities and things we eat, we also have a signature frequency. We hypothesize that this frequency does not readily change and is electromagnetic in nature, vibrating around our physical electric beings. This is based on three important factors:

1. Research has indicated that we have a unique scent (i.e., not the scent that is impacted by the garlic we eat) that does not readily change. Since scent is a vibration, we at Phia Lab call this our signature frequency.
2. We are electric beings. The average human, at rest, produces around 100 watts of power.[41]
3. Through Maxwell's equation, we know that magnetism always operates in a loop around an electric charge.

If we can match the electromagnetic energy vibrating around our electric beings or our "signature" frequency in a constructive way to the electromagnetic energy of biological essences, the energy body can be amplified, untangled and better aligned.

---

41. Robert, Becker, Gary Selden The Body Electric: Electromagnetism And The Foundation Of Life. 1st Edition. (William Morrow Paperbacks, 1998).

At Phia Lab, we developed a seven-page questionnaire to determine this "signature" frequency for hundreds of test subjects. Test subjects were instructed to fill out the questionnaire. Then they selected from the 60 to 80 essences evaluated in this research. The essences selected were narrowed down to eight. The eight essences were blended into three sample variations. The test subjects were then told to take the three samples home and try them on one at a time to see how they reacted with their body chemistry.

Once the subjects had some time to interact with their samples, they selected their favorite variation. The electromagnetic energy of their final selection was calculated and compared to the results of their survey.

Through extensive evaluation of the data collected, we were able to build an algorithm. This formula tied specific answers from the questionnaire to a certain frequency. Once the individual's frequency was calculated we were able to prescribe two of the six Energy Waves as a possible fit. Compilations of all the essences found in each Energy Wave were created and bottled. One drop from the bottle was applied to the individual's pulse points and results were recorded.

We found that for people not going through a major life change, the algorithm predicted an Energy Wave of essences that matched the compilation of essences they liked the most if they were smelling them directly 95 percent of the time. If the person was going through a major life change, either positive or negative, the formula predicted a compilation of essences to the right or left of the preferred Energy Wave if the subject was smelling them directly 75 percent of the time. This was attributed to the fact that if a person is going through a major life change – such as a divorce, a death in the family, a move, a new baby or a new job – their energy shifts to entrain with the change. This tends to shift their senses, most notably their sense of smell. In the cases of a major life change, the energy compilation liked the most when directly smelling them tended to be a poor match to their signature frequency.

From this research, a more condensed version of the seven-page questionnaire was developed. This allowed for a person to answer 13 questions that could predict their signature frequency and then prescribe an Energy Wave of essences that would resonate with them most effectively and therefore be a good fit. We call this condensed version of the original questionnaire the "energy fit guide." We have found this profile to be very effective at predicting the Energy Wave that best aligns with the person. When evaluating the person first with the energy profile and then with kinesiology and biofeedback to verify results, we have found the profile to be over 90 percent accurate. You can find the energy fit guide on the Phia Lab web site (www.PhiaLab.com) and try it for yourself.

Biofeedback

To further test the ideas and conclusions gathered from the research, an advanced biofeedback device was acquired. The device reads body temperature, heart rate, galvanic skin response (the emotional arousal obtained by recording momentary changes in perspiration and other functions of the autonomic nervous system), and acupuncture points in the hand. This information is fed into a third-party algorithm that produces information on the energies and emotional state of the person.

The biofeedback device was constructed by Dr. Holebreck, one of the original developers of the lie detector and biofeedback devices. The software and algorithm are the results of over 40 years of research by various doctors and alternative medicine specialists. The system is designed to gauge the energy around each of the energy centers in the body.

Subjects were asked to fill out the condensed energy profile and then interact with just one essence from their recommended Energy Wave. In these cases, the subjects demonstrated a small shift in their energy readings. For this reason, we decided to test the compilation containing all the essences in each Energy Wave to see if the effect was amplified. They would then interact/apply one drop of the compilation of essences from their recommended Energy Wave.

Over 75 percent of the time, the shift in their overall energy readings via the biofeedback device showed an increase of more than 40 percent. This can be seen via the circular diagram in the bottom right corner of the readout in Figure 7. This diagram represents an average energy reading of the energy centers in the body in terms of their intensity and balance with each other. When interacting with a

Figure 7

Energy State Before and After Exposure to Constructive Energy Wave Compilation

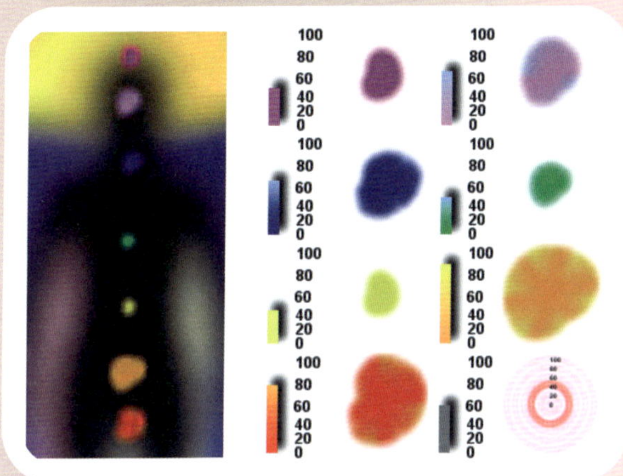

Before Exposure

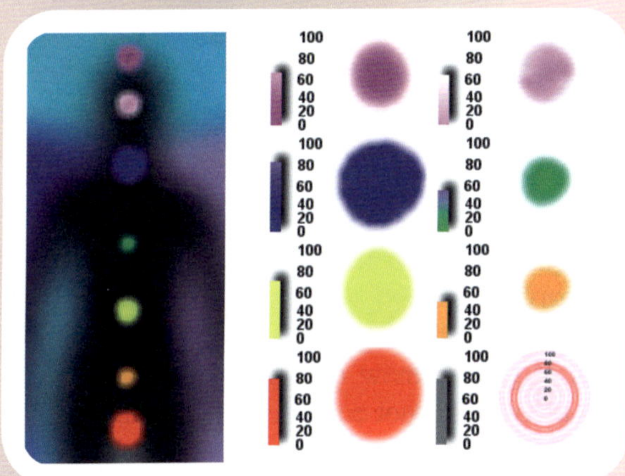

After Exposure

compilation from an Energy Wave not recommended by the energy profile, there was little to no detectable shift in their energy reading.

When essences are combined according to their energy fingerprint, they have a noticeable effect on a person's energy.

You can see on the readout that the energy centers aligned and balanced out in shape, size, and color after interaction with the compilation of essences from the recommended Energy Wave. This has been a common trend in all the test subjects evaluated to date.

This research has been successful in demonstrating that these essences, when combined according to their energy fingerprint, have a noticeable effect on a person's energy. This requires that the Energy Wave used is constructive to one's signature frequency. There is little effect on one's energy if only one essence within the recommended Energy Wave is used. There is no effect or a slightly negative one when an Energy Wave compilation is used on a test subject that is not in a constructive pattern to the test subject's signature frequency.

The biofeedback device has been useful in testing the accuracy of the energy profile's predicted best fit energy compilation of essences. In 80 percent of all case studies so far, the predicted Energy Wave has increased the overall energy reading in terms of intensity, clarity and balance by 30 percent or more for the person smelling and interacting with it.

More research is needed here, but these initial results point to some interesting findings that seem to be a means of further verifying the hypothesis of this research.

Summary of Findings

Traditional aromatherapy methods are such that if a person is having trouble sleeping, they may prescribe some lavender essence. If that person really likes the smell of juniper, they may add a little juniper essence. Our research indicates that this is counterproductive because lavender and juniper have destructive frequencies. This means that when they are combined, they interact in a destructive pattern dissipating the effect of each essence on its own. Furthermore, lavender may be good for some people if it constructively matches their signature frequency, but not effective if it matches in a destructive way. Similarly, juniper would be beneficial if its frequency is a good fit to the user's signature frequency, but not all that effective if it is not a good frequency fit. This is counter to what aromatherapy practice and belief has been and is our new science.

At Phia Lab, we combine essences based on their energy fingerprint, thereby amplifying their energy and the effect they can have on the one using them. This effect increases exponentially when the frequency contained in the bottle is in a constructive pattern to the signature frequency of the one using them. To simplify the concept of combining essences, think of how music is created on the piano. Certain keys can be played together to create harmony. Add a key with a different vibration, and you will hear dissonance. In essence, we are harmonizing essential oils by measuring their vibrations.

This research helps explain why the essences used in my "scent of competition" were so helpful in clearing out tangles and blockage in my energy body during my World Cup years. I was using essences with a similar energy fingerprint and blending them together without a full understanding of it at the time. The energy in the blend was amplified as a result of constructive interference patterns between the essences. I also blended everything in phi proportions or sacred geometry which I believe helped better synchronize the energy. As it turns out this scent of competition blend was a very close match to my signature frequency. For this reason, it was effective at untangling, aligning and amplifying my energy body. Being free and clear of any obstructions in my energy body helped me transform any negative thoughts into positive ones quickly. I could do it quickly because nothing was blocking the emotion tied to the more positive thoughts. Hence shifting my energy and gaining momentum in a positive direction took little effort. This also explains why my blend did not have the same effect on everybody who tried it. If the person trying it had a similar energy to mine it worked very well, but if

> The energy in my "scent of competition" was amplified as a result of constructive patterns between the essences.

their energy was in a destructive pattern there was no noticeable effect.

It is my hypothesis that people who maintain a clean and clear frequency prevent the development of symptoms and illnesses. Of course, in practice this does not work for most of us because we face a steady stream of obstacles. We experience stress and emotional challenges daily, which tangle, cloud and block our bodies frequency. Hence, we need to cleanse and strengthen our body's frequency regularly/daily with the right substances that are compatible at the cellular/energetic level of our being, rather than wait until our frequency has become tangled and murky and a friendly host for microscopic invaders. In practice, this form of quantum nutrition has worked time and time again in cleansing and illuminating my energy and the energy of my family.

It is exciting to think about the possibilities of connecting these dots of progress with more advanced research that extends the work herein. It is difficult to reject the view that continued research in this area will deliver positive impacts on the health and wellness of humankind.

Our specific research into biological essences was able to establish from a scientific perspective:

1. Biological essences have electromagnetic energy that is vibrant, varying and quantifiable.
2. These varying energies can affect our emotions and therefore, our energy.
3. Combining biological essences of a similar energy fingerprint amplifies their effect on the people interacting with them.
4. If we can match a person's signature frequency constructively to the frequency in specific complications of botanicals, a positive effect on their energy body can be detected.

New scientific research is pointing to a correlation between the energy humans emit and their emotional state. Furthermore, science has proven a link between our emotional state and our physical health, not to mention thousands of years of Ayurvedic and traditional Chinese medicine principles that assert the same thing.

Dr. Elizabeth Blackburn's work shows how meditation (focused thought and emotion) changes the telomeres found at the ends of our

An Energy Wave matched to one's signature frequency is quantum nutrition for the energy body.

chromosomes resulting in a reversed aging process. Dr. Blackburn literally demonstrates how focused thought and emotion can change our physiology.[42] In this context, one can conclude that these biological essences affect our energy and therefore our emotions and, as a result, our physical health and wellness.

You can think of a Phia Energy Wave that correlates to one's own signature frequency constructively as medicine or nutrients for one's energy body, or quantum nutrition for a whole- body approach to wellness.

These findings shed light on a grey area in science as well as an area that has been excluded from the mainstream teachings in Newtonian medicine. With advancements in measurement, approach, and technology, the potential impacts of this line of research could be far-reaching.

---

42. Elizabeth Blackburn, Accelerated telomere shortening in response to life stress (Proceedings of The National Academy of Sciences, 2004).

Implications and the Broader Context

Despite the fact that plant-based ecosystems have evolved over billions of years, we are just beginning to understand better the survival attributes that plants have perfected over time. The biological essences of plants and their energy vibration are essential to this process and this area of study is gaining some momentum. In addition to the research herein, other developments are contributing to advancement in this field of energy.

To provide some context to the discoveries made in the lab as part of this effort, it is important to point out some examples of how progress is being made in related areas. This may inspire more creative thinking on how to bridge various gaps on the cutting edge of this research.

1. Healing — Reiki is a healing modality from Japan developed in 1922 that deals with the energy surrounding us. The now-famous cardiovascular surgeon Dr. Mehmet Oz brought tremendous attention to Reiki when he invited Reiki practitioners to treat patients during open-heart surgeries and heart transplant operations. Dr. Oz is often quoted as

saying, "Reiki has become a sought-after healing art among patients and mainstream medical professionals." Reiki is now used in more than 800 American hospitals as a means to accelerate the healing process and to alleviate pain.[43] If hospitals are starting to understand that the energy around us affects our physical health, then it follows that the energy from biological essences from the botanical world carry tremendous healing abilities.

2. Therapeutic Forests — There are 24 certified therapeutic forests in Japan. Within these forests they have proven, by measuring heart rate, body temperature, brain activity and blood pressure, that the forests are having a positive effect on the health and wellness of humans experiencing them.[44] In this research, people have a measurable physiological reaction to their environment. By extension, the Phia Lab research supports that the high vibration from botanicals

---

43. Baldwin AL, Vitale A, Brownell E, Sicinski J, Kearns M, Rand W. The Touchstone Process: an ongoing critical evaluation of Reiki in the scientific literature (Holist Nurs Pract, 2010).

44. Park, B., Tsunetsugu, Y., Kasetani, T., Kagawa, T., and Miyazaki, Y. The physiological effects of Shinrin-yoku (taking in the forest atmosphere or forest bathing): evidence from field experiments in 24 forests across Japan. Environmental health and preventive medicine (2010).

are major contributors to the measured positive health and wellness impacts.

3. The Wim Hof Method - Wim Hof, also known as "The Ice Man," has a scientifically proven method for literally changing one's own physiology via breathing techniques, extreme cold temperatures, interacting with nature and meditation. His method demonstrates how important our breath is to regulating body functions and the positive effects of nature on our mental health.[45]
4. Paying for Wellness - The Cleveland Clinic conducted a study where they paid employees for five wellness behaviors: eating healthy, getting eight hours of sleep, working out, meditating and interacting with nature. According to Michael Roizen, MD, chief wellness officer at the Cleveland Clinic, over a five-year period, they were able to show millions of dollars being saved in medical costs and sick days.

45. Wim Hof, The Way of the Iceman: How the Wim Hof Method Creates Radiant, Longterm Health (Dragon Door Publishing, 2017).

Chapter 8 The Zone

Perhaps the most common word associated with elevated performance, the "zone" is where the magic happens. Many perspectives have been shared on this elusive state of mind by athletes, coaches, psychologists and perhaps a few scientists. The spectrum of opinion on the path to reach this highly sought-after place is wide. However, most agree that elements of the zone drive the ability to repeat and sustain heightened levels of performance. You don't need Olympic aspirations to benefit from this thought framework. It applies to all walks of life across a variety of situations from special events to daily activities and in interactions that you may have from the adult stage to your child's playground. For most of us, it requires more than flicking the switch to on. But the benefits of finding your path to get there are limitless.

In this chapter, I share my views on the zone based on my unique experiences as an athlete and a scientist. This perspective originated with two divergent and youthful goals identified when I was five years old. These goals have unexpectedly converged. The outcome is a science-based, field-tested alternative to the more traditional viewpoints on how this magic is created. Open your mind and let your curiosity wander.

Complete Coherence

As elite athletes on the U.S. Ski Team, we worked hard to evoke the "zone" on competition day. On the mogul course, the zone is the feeling, emotion, and memory of skiing to perfection. It is the state where there is no misstep, where everything flows smoothly, and things fall perfectly into place. When in the zone, everything acts as one unit. It is a state that equates to being completely present in that moment with organized information (coherence) between the mind, physical body, and energy body. Some people may call this flow.

A Scientific Perspective

From a scientific perspective, the state of being in the zone defies the Second Law of Thermodynamics. The Second Law states the entropy (disorder) of an isolated system will always increase over time. In other words, the zone equates to a decrease in entropy (i.e., a more orderly system).

I agree with Nassim Haramein, a physicist specializing in unified physics theory who founded the Resonance Science Foundation, and said, "If more scientists spent more time doing sports outdoors it would help them understand the natural world around them." From my experience, if the system for a single person is defined as the mind, the physical body and the energy body, I would argue that entropy decreases when one is in the zone. This entropy decrease is the state elite athletes are trying to attain.

On competition day we would all try to summon the zone through our senses. Since our sense of hearing has a direct link to our nervous system (Dr. Andrew Weil), we all played a certain type of music on competition days to help calm nerves and evoke the zone state. We would utilize our sense of sight by visualizing ourselves skiing perfectly. We would wear a certain pair of "lucky socks" for a certain type of feel. We would also have a specific drink or food on competition day. But most were overlooking the sense of smell, the sense that is directly related to the part of the brain we were trying to tap into, that emotion and memory of skiing perfectly.

Our sense of smell is our one sense directly tied to our emotions and therefore our energy body. The most effective and quickest way to achieve coherence between the energy body and the physical body is through our sense of smell. But what smell or frequency will evoke greater coherence between the two bodies?

In our research, we found this task to be a bit like finding the right radio station that plays all your favorite music all day long, only better. This is better because we believe certain botanical essences are made by nature to provide greater coherence in the energy body, the physical body and mind.

If we want to determine which frequency from botanicals will assist us in becoming more coherent between our bodies and mind, we must once again look to Einstein's natural law of

$E=mc^2$. This law tells us everything is energy and that physical matter is just energy slowed down enough for our senses to perceive it. If this is true, it is important to look at everything from the perspective of energy, frequency, and waves.

When looking at everything from this perspective it is useful to consider constructive and destructive frequencies like those discussed earlier. If we can constructively match our own signature frequency to the frequency in specific biological essence compilations, we can determine what smell or frequency will increase coherence between our physical body, energy body and mind. This will decrease entropy and thus help to evoke the zone.

Taking it a step further, the zone is a state of decreased entropy and increased Gibbs free energy. For the past 150 years, entropy has been a very important contributor to Gibbs thermodynamic free energy function, G, given by:

$$G=PV+E-TS$$

Where	E=energy,
P=pressure,	T=temperature and
V=volume,	S=entropy.

Nurturing my energy on a regular basis with the right frequency found in biological essences helped me get in the zone and stay there.

Changes in Gibbs, G, drive all the processes in nature that have so far been discovered.[47] Gibbs free energy dictates all things in the macrocosms and microcosms. If you want to change the world you have to change the free energy all around you. One person can do this by bringing their body into greater coherence or, as we athletes like to call it, by evoking the "zone."

When in the zone, you are not in the fight or flight or reactive state of the subconscious mind. Nor are you in the meditative state of the creative conscious mind. Rather, you are in the sweet spot right in-between. From this place, incredible things are possible. When I was in the zone, I always had to remind myself to stop thinking. When you are thinking, you can get pulled into the conscious

47. William Tiller, Ph.D., Walter E., Jr., Dibble, Ph.D., White Paper I - A Brief Introduction to Intention-Host Device Research (Tiller Institute, 2009).

mind and are no longer in the sweet spot or zone state where mind, body and energy are coherent. To maintain the zone state, there was no room for fear or ego to exist. From this place, I would ski fully exposed with my heart on my sleeve, and this is where remarkable maneuvers and astonishing accomplishments came from.

To look at this from another broader perspective, according to Dr. Tiller, if we could bring all the electrons in a light bulb into coherence, we would generate the amount of light coming from the sun. Just think what could happen if all the electrons (about 1.5 x 10^28) in your body were in coherence ... you would truly be a force! Now imagine what one million, or for that matter, one billion people operating in their zone could do. If we can let go of our egos, release fear and become more aware of the energies around and inside us, we could move mountains!

Sustaining The Zone

Reaching your zone is not hard. You can use all your senses to evoke the zone: seeing, feeling, hearing, tasting and smelling, but you must also recognize that there is energy all around you that can help or inhibit you. Learn how to utilize the following natural laws:

1. The Law of Entanglement or Entrainment – Protect yourself from energies not serving you by surrounding yourself with a "good energy team" and avoiding people, places, and situations that carry energy not serving you. By the law of entrainment, you do not want to be entwined with energy destructive to yours. On the flip side, you want to entrain with energy that is constructive to your own and, thus, help elevate your frequency. This is not always possible, but it is important to recognize the toll it can have on your energy body and have a plan to replenish it.
2. The Law of Attraction – It is not just the thought of the reality you want, but the frequency that allows this law to be utilized. In other words, you have to think about the reality you want, but also feel what it would feel like to have that reality. By the law of Attraction (physics, not philosophy) you get what you put out.
3. The Law of Conservation of Energy – Energy cannot be created or destroyed, and that includes the energy in your thoughts.

You have to transform negative thoughts into something beneficial. You can not ignore them or they will obstruct the emotions tied to the thoughts of the reality you want.

4. Energy = Mass x Conversion – Sometimes we carry around things in our energy body that we are not even aware of. These things can hold us back from reaching our highest potential. After you understand that we all have an energy body and it is far bigger than our physical body, you can also find ways to nurture it.

Utilizing your senses and these four natural laws will help you get in the zone and stay there. From a state of being in the zone:

1. You avoid entanglement with energies not serving you by being the stronger system, i.e. other energies will entrain with you instead of the opposite taking place.
2. You radiate a clearer, more intense frequency; therefore, you will attract a clearer, more intense frequency.
3. You bring greater coherence to your physical body which is tied to increased health, wellness and performance.

Sometimes using my senses and applying the four natural laws in real-life situations to evoke the zone was easier said than done. I found great assistance in nature, more specifically energy nutrients found in nature's botanicals. Nurturing my energy body on a regular basis with the right Energy Wave truly helped me be more coherent, vibrate at the frequency of the reality I wanted and be the stronger system. This can work for you as well.

Back in the 1970s, researchers proved, via solid science, that plants can read our emotions – in other words, our energy. Findings from various research studies including our own indicate that our emotions are tied to our energy bodies. Our research demonstrated that essences from botanicals can help cleanse, amplify and align our energy bodies if we interact with the essence compilation with a frequency that is constructive to our unique signature frequency. In other words, biological essences can help transcend the limitations we have imposed on ourselves.

Olympians, elite athletes, performers, and executives are always trying to evoke the zone on their "game" day. This means that they are striving for coherence between mind, body (all 50 trillion cells) and energy to be at their best. Our sense of smell is directly related to the part of the brain associated with emotion and memory. Emotions directly affect our energy body. Our energy body or the frequency we emit has a direct effect on our physical performance and our health, thereby determining the life we live.

"You just have to dive in and learn to unlearn what you have already learned." ~ Malcolm Gladwell

You just have to dive in and "learn to unlearn what you have already learned," as Malcolm Gladwell said in his book Outliers. By replacing the limiting beliefs that you've learned from your families, friends and social institutions with a new self-empowering perspective, you can open your eyes to unlimited potential. In doing so, you can control your physiology, your destiny and create your reality ... and you don't have to be an elite level performer to do it.

You can be more effective in your day, have more energy at the end of the day and even have more restful sleep by evoking these four natural laws and using nature as your personal assistant.

You decide if you want to run at full speed, expand, elevate, love wastefully, be passionate and live fully. The choice is yours to make, and nature is there to support and help you with whatever path you choose.

On a much broader scale, society is at a shifting point from the old human story of separation, compartmentalizing and Newtonian thinking to a new story of connection, cooperation, and Einsteinian thinking. Collectively, we are in the zone, a rare sweet spot between old and new ways of thinking, where we all have the power to do astonishing, unthinkable things and literally be a force in all we do!

200 ml
±5%
50
150
100
150
50

Chapter 9
A New Start Gate

After seven years of competitive skating and 16 years of skiing on the World Cup circuit, I now find myself in a different kind of "start gate." In this one, the goal is not delivering a perfect 30-second physical performance of turns, speed and air. It is the start of a longer race to make the biggest and best impact. This is about helping people reach and sustain their zone as individuals and as communities. I have barely pushed out of the gate on this and have an incredible amount of work, discovery and fun ahead.

I believe that the path to reach this destination centers on three main principles:

1. Accept that everything is energy and re-examine opposing beliefs, specifically in medicine.
2. Consider thought as one of the most powerful forces in the universe; and
3. Adopt a more informed and comprehensive health and wellness approach to life.

In line with these guideposts, I am working to expand this thinking through bioenergetic research and provide people with the opportunity to experience the benefits of biological essences through the use of wellness products we create at Phia Lab.

Newtonian Vs. Einsteinian Views on Medicine

It has been over 100 years since Albert Einstein's Theory of Relativity and the introduction of his famous equation $E=mc^2$.[48] Scientists have known for some time that matter can change form into energy and energy into matter. Still we have resisted working this natural

48. Walter Isaacson, Einstein: His Life and Universe, Simon & Schuster Paperbacks, (New York, 2008).

law into modern allopathic medicine[49] or our approach to health in general. This has given rise to two separate views on medicine. One view that is considered allopathic and the predominate practice in most developed countries. The other view that is considered alternative with its roots in ancient Ayurveda and Chinese medicine.

Allopathic medicine practiced today is rooted in 17th-century science or Isaac Newton's (Newtonian) principles. If an individual has a problem, traditional medical practice is to cut out the affected area. If that does not work, doctors are taught to poison the area. The medical education system is slowly beginning to adopt new principles. However, the core curriculum continues to focus on methods rooted in a time before we knew energy could change form to matter and vice versa. As a result, it makes sense that allopathic doctors do not recognize these biological essences as having a true medicinal effect because they fail to see the energy in the physical matter of botanical essences.

For the most part, modern medicinal principles ignore the broader system and focus on a single area of activity. If you go to the hospital, you are

---

49. Richard, M.D. Gerber, Vibrational Medicine, (Bear & Company, Vermont, 2001).

more or less "processed." You fill out the form, and then you see the cardiologist who specializes in the heart, then you see the otolaryngologist who specializes in the ears, nose, and throat, etc. You might see an orthopedic doctor to repair trauma in your knee. However, the orthopedic doctor will most likely not be prepared to talk about a heart issue or how your mind impacts healing because these areas are outside the boundaries of his or her expertise.

Allopathic medicine analyzes the body in parts or components and misses the whole picture. Recall our discussion about myrrh essential oil. Analyzed in its components, it is toxic, but as a whole, it is quite medicinal. By failing to see the whole system, we are more likely to miss the beautiful healing qualities of the human body and nature.

From the perspective of allopathic medicine, more is usually better. For example, synthesized drugs are concentrated with specific molecules packed into a pill or another form of matter. Ask your doctor at your next checkup if there is a homeopathic alternative to any medication he or she prescribes. The response is likely to be that the potency of homeopathic remedies is far too small to have any effect on the human body.[50] This

50. Richard, M.D. Gerber, Vibrational Medicine.

is because allopathic medicine fails to recognize the energy component in medicine.

It is disappointing when a pharmacologist cannot recommend the use of lavender for a burn because scientists/researchers cannot find an active ingredient in lavender that mimics the effect of the whole oil. Numerous studies of essential oils point to lavender as an effective burn healer. According to Healthline.com, "it has pain-relieving properties, the ability to reduce inflammation, and antimicrobial activity." Could it be that the mechanisms for how these essences work cannot be explained via the standard scientifically assertible reductionist argument, so their healing abilities are kept out of mainstream medicine?

In contrast, alternative medicine or an Einsteinian approach views the system as a whole and not the sum of its parts. An alternative medicine approach also views a lighter concentration of molecules with an energetic vibrational pattern as most effective.[51] Practitioners of alternative medicine based on ancient practices or Einsteinian principles view everything as energy. Herbalists take small concentrations of herbs and prepare them in

---

51. Richard, M.D. Gerber, Vibrational Medicine.

a way that the vibrational energy is absorbed into water or powder form. Flower essences for healing are prepared similarly, but they are known to have the highest vibrational frequency and, therefore, the widest range of physical and mental conditions that they can affect.[52]

It is interesting to consider what experts in holistic medicine have to say. Norm Shealy, M.D., founding president of the American Holistic Medical Association, has flatly stated that "energy medicine is the future of all medicine." Donna Eden writes in *Energy Medicine: Balancing Your Body's Energies for Optimal Health, Joy, and Vitality:* "When all your energies are brought into harmony, your body flourishes. And when your body flourishes, your soul has a soil in which it can blossom in the world. These are the ultimate reasons for energy medicine—to prepare the soil and nurture the blossom."

A good example of energy medicine is captured in one of Dr. Tiller's white papers as noted below.[53]

If one takes a beaker of water with some bacteria in it and then shakes some silver (Ag)

52. Gurudas, Flower Essences and Vibrational Healing (Cassandra Press, California, 1989).
53. William Tiller, Ph.D., White Paper XXIV - An Introduction to Psychoenergetic Science & Information Medicine.

colloid particles into the water, we all know that the bacteria will probably be killed via the bactericidal action of the Ag particles. The general conclusion drawn from this observation is that the physical contact between Ag and the bacterium is a necessary condition for the killing of bacteria. This, in turn, has led to the assumption that pharmaceuticals do their work in the human body via contact-types of chemical reactions and this has led to what is labeled as today's chemical medicine.

What most people do not know is that, if one takes a fluorescent tube held horizontal and places silver (Ag) colloid particles in it and then focuses the output light from the ignited tube onto the beaker of water containing bacteria, one also kills the bacteria. Such an experiment shows that it is not the physical contact between Ag and the bacterium that is necessary for the killing process to occur. Rather, it is one or more different types of photons from the electromagnetic (EM) emission spectrum of Ag that entangle with the EM carrier wave from the fluorescent tube and are transported to the beaker of water that are the actual killing mechanism involved in the demise of the bacteria.

You may want to read these paragraphs again if a feeling of wow has not consumed your mind!

Today we see matter or chemical medicine, not energy medicine, but this perception is shifting. More scholarly research is pointing in this direction. Dr. Blackburn is just one example of someone who has done remarkable research showing how focused thought energy can change our physiology.

Energy-based healing has also been used to explain the placebo effect. This remarkable phenomenon occurs when a placebo – a fake treatment like a sugar pill – improves a patient's condition simply because the person has the expectation that it will be helpful. In a study at the Baylor School of Medicine, surgical options were sought for patients with debilitating knee pain. Patients were divided into three groups – one group had the damaged cartilage shaved from their knee, the second group had the damaged tissue flushed from the knee joint, and the last group was sedated, had saltwater splashed into their knee, and was told that invasive surgery had been performed when in fact it hadn't. Amazingly, the group that received no surgery improved as much as those that had.

Typically when testing a new pharmaceutical in a lab for its effectiveness, if it does not outperform the placebo effect, it is considered a failure. However, we are finally starting to recognize the placebo as a contributor to the effectiveness of any drug. Alia Crum, Ph.D., assistant professor of Psychology at Stanford University, states that they are looking at how they can build on the placebos powerful effect with the use of medications.

As I have shared, I had the pleasant experience of having six knee surgeries. Soon after the sixth surgery, doctors recommended a seventh to repair cartilage and scar tissue. This time, I said no. Believe it or not, at this point in my career I had learned how to manifest things into reality, so I healed my knee with the energy from my own thoughts. I proceeded to ski another seven years on the World Cup circuit injury and pain-free, and continue to have no pain till this day.

I guess you could say that I benefitted from the placebo effect, mind over matter. Recent studies indicate that this placebo effect is up to about 80 percent effective. This is just one example out of many on how we can consciously affect the biology of our bodies by the energy in our thoughts, so long as our thoughts are clear, focused, and tied to a strong emotion.

Liberated Thought

In science today there are four fundamental forces of nature: gravity, electromagnetism, strong nuclear force and weak nuclear force. It is my view that there is a fifth force not readily recognized but which is stronger than all the rest. This fifth force is thought or, more specifically, "liberated thought." I define liberated thought as the thought that comes from a place of being in the zone, i.e., highly intense positive thought tied to intense positive emotion.

Think about it. You overcome gravity by sending a thought of lifting your leg to the muscles in your leg that lift your foot off the ground. Through the ages there have been reports of people reversing the gravitational pull to levitate via liberated thought. Electromagnetism is the force that holds atoms together, and electromagnetic interactions are responsible for generating the light emitting from your light bulbs.

I have heard of people with liberated thought being capable of making lights flicker on and off. The remaining two forces work at the atomic level. The strong nuclear force holds the nucleus together. The weak nuclear force is responsible for radioactive decay. Both Dr. Emoto and Dr. Tiller demonstrated that the molecular structure of water can be altered with liberated thought.

These discoveries are forcing scientists to research the effect of thought and unseen energies as having real health and wellness qualities in today's society. The Global Wellness Institute has created www.wellnessevidence.com, an online platform where peer-reviewed research in the area of holistic health is emerging at a steady pace. However, there is still a serious need for more scholarly research in this area.

Coming from a hard science background, I am consistently surprised at the lack of data. I have also been delighted to see it steadily grow in the last ten years. Many thanks to people like Dr. Elizabeth Blackburn, Dr. William Tiller, Dr. Bruce Lipton and many more who have had the confidence, courage and passion to step outside the traditional structure of academia to do something that has been classified as taboo for so long. Their work is revolutionary and continues to pave the way for new science and new physics to shape the reality of the 21st century. They are leading the path for the alchemists of the future.

You can evoke the alchemist in you now. Recognize that *liberated thought* influences the energy around you. The energy around you is what makes up the mass that you perceive through your senses. In doing so, you will have more control over your health, wellness, and physical performance. You can tap into the unseen energies and unlock potential you never knew you had.

To explain how energy is transformed into

> The unseen energy all around you is what makes up the mass your senses perceive. You can tap into these energies and unlock potential you never knew you had.

mass from a more scientific perspective consider the Double-Slit experiment first performed with light by Thomas Young in 1801. In 1927, Davisson and Germer demonstrated that electrons show the same behavior, which was later extended to atoms and molecules.[54] In this experiment electrons are fired through a wall with two slits and then recorded where they land on the screen behind the wall, Figure 8. The expectation was that the electrons would form a pattern on the screen similar to the two slits in the wall they were passing through. However, this did not happen. Instead, the

Figure 8
Double-Slit experiment

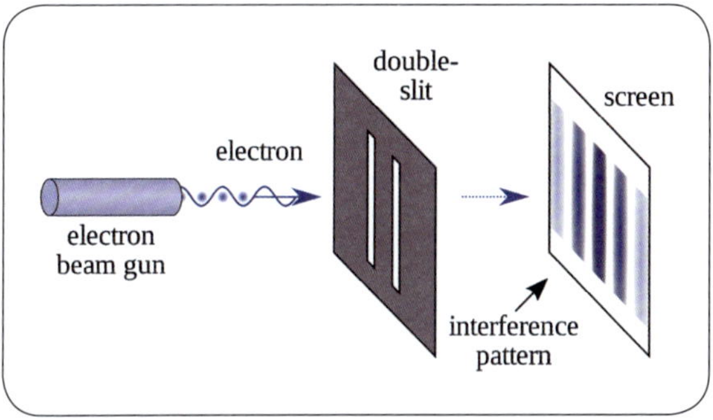

54. "Physicists Smash Record For Wave-Particle Duality"; Eibenberger, Sandra; et al. "Matter-wave interference with particles selected from a molecular library with masses exceeding 10000 amu". Physical Chemistry Chemical Physics (2013).

electrons formed a pattern on the screen that was indicative of a wave as opposed to a particle. It gets even more interesting when you consider the fact that once a camera was placed at the location of the double-slit wall the electrons formed a pattern on the screen indicative of a particle. The astonishing conclusion was that electrons are not particles, but rather waves or energy until they are observed. Schrödinger derived the Wave Function equation so that at any given time (t), you get a value Ψ (x,y,z,t) for all possible locations (x,y,z).*
Schrödinger was the first to write a wave equation to describe the behavior of a particle.

In 1926 the physicist Max Born came up with a probabilistic interpretation. He postulated that the square of the absolute value of Schrödinger Wave Function |Ψ(x,y,z,t)|^2 gives you the probability density for finding the particle at position (x,y,z) at time (t). In other words, the probability of an electron going through one of the slits and hitting the screen behind it in a specific spot. Max Born's equation gives you a probability of where you might find a particular electron.

---

* $$\frac{ih}{2\pi}\frac{\partial\Psi}{\partial t} = -\frac{h^2}{8\pi^2 m}\left(\frac{\partial^2\Psi}{\partial x^2} + \frac{\partial^2\Psi}{\partial y^2} + \frac{\partial^2\Psi}{\partial z^2}\right) + V\Psi.$$

Here V is the potential energy of the particle (a function of x, y, z and t), i=sqrt {-1}, m is the mass of the particle and h is Planck's constant. The solution to this equation is the wave function Ψ (x,y,z,t).

So how does the wave function collapse? Scientists will tell you nobody really knows. I believe it is the act of our own thoughts that collapses the wave function and brings things into our reality or energy into matter. We are not victims to our environment, our genes or our circumstances. We are in control of it all through our thoughts or rather *liberated thoughts*. The deeper you go in nature the more powerful nature becomes, i.e., nuclear power is a million times more powerful then chemical energy. Similarly, the deeper and more focused your thoughts the more powerful they become.

A Refined Approach to Healing

Several years ago, I heard a thought leader in wellness say that spas are going to start replacing hospitals in 10 to 20 years. At the time, I thought that sounded great, but I had my doubts. I had always seen spas and hospitals as completely separate. Spas have been considered a destination or a luxury experience when you want to treat yourself. A hospital is where you go when you are sick or have trauma. Now I see how conducive spas are to our health and wellness. Going to a spa on a regular basis helps maintain health and wellness, but it can also help you heal.

I understand first-hand that the path of living a full and passionate life can occasionally result in a physical setback or two, as described in prior chapters during my skiing career. Fortunately, none of those injuries were life-threatening. However, since retiring from the World Cup circuit, I have faced two near-death bike "experiences" (i.e., accidents). The first occurred on a paved bike path in Denver. I was on a road bike attempting to pass my husband when someone stepped onto the path in my lane. I pulled back behind my husband's bike to avoid the pedestrian, but accidentally clipped his back tire and flew into the rough grass and rocks along the path. The bike flipped at a speed of roughly 25 mph and I landed on a large rock with most the impact on the right side of my abdomen. Unknown at the time, it was diagnosed in the emergency room as a a stage five liver laceration. I was bleeding internally and nearly bled to death. I believe that my ability to vibrate at a high enough frequency enabled my body to clot up the wound just in time. Instead of a major surgery, specialists opted to insert a stint. I was home after three weeks in the ICU.

The second experience occurred more recently around the time I was finishing this book. I was riding a mountain bike on a freshly re-graveled dirt road from my home to the trailhead near Winter Park. I saw an oncoming car come around the turn in the distance. As I moved over to the side of the road, my bike slipped on the loose gravel and I went down. As I hit the ground I tried to get out of the way, but I was having trouble getting out of my clip pedals. I kept saying to myself "he is going to stop before he gets to me," but he didn't. When the car struck, I was trying to hold on to something under the front bumper to stay ahead of the driver's front left tire. My left leg got caught under the car and bore the brunt of the impact breaking my femur in two places. The car pushed me up the hill under it for about 11 feet. I was trapped under the front of the car and lost a lot of blood and skin from being pushed/dragged along the gravel road.

I truly believe that my experiences utilizing the laws of physics laid out in this book and harnessing the bioenergetics in nature are what saved me, not just once but twice! But don't take it from me. According to the Navy SEAL Training Guide, "the most important survival trait is — the ability to remain mentally focused when faced with continuous and very stressful challenges or situations."

After spending over a month in the ICU with this "experience," I can say that hospitals today are great for trauma. When you have multiple broken bones or other severe injuries, the last thing you want is a massage or acupuncture. You need a surgeon! However, after the surgery, you need a beautiful, peaceful place to recover.

Hospitals can be so focused on illness or injury that the wellness component is overlooked. Imagine if hospitals were built with more focus on wellness – an environment that is by all means sterile, but more uplifting. Recall if you can't match the frequency of the reality you want, you do not get that reality. If you are stuck in a hospital with no color, low energy and a feeling of loneliness or despair, how does that promote healing in the physical body?

Picture this for one moment. You check into a hospital for a procedure and fill out a five-minute profile that determines your signature frequency.

You then proceed to a room that is either painted or decorated with colorful art in a color constructive to your frequency. Beautiful nature videos and soothing sounds are coming from the television, and the room is being diffused with delicate biological essences customized to your energy. The bathroom is stocked with effective, but all-natural body products that contain delicate bioenergetics from nature in a constructive frequency to your own. Instead of long grey and white walls and doors, hallways and walkways are filled with nature sounds and feel a bit more like an adult Disneyland.

Maybe it is not spas replacing hospitals, but rather a merging of the two to truly provide health and wellness to people in the 21st century.

However, if people would start to embrace spas as a source of wellness and go regularly basis, then the need for hospitals would decrease. Studies show 70 to 90 percent of all illnesses are preventable with regular wellness practices.[55] Like Sri Swami Satchidananda has stated, replace the "i" in illness with "we" and you have wellness!

I see this happening around the world as I continue to work with spas to incorporate the Phia energy products into their treatments. In doing so,

55. B. Lipton, The Honeymoon Effect.

we hope to make an impact in bringing energy medicine to the mainstream, helping everybody live their best life.

Phia Lab

At Phia Lab, we harness energies created by sun, soil, and water found in fields, forests and oceans from the far corners of the world and combine them using the natural laws of physics and mathematics. Ultimately, we create energy products that help us all live fully and with passion – to be a force!

Similar to the energy medicine example with the silver colloid particles killing the bacteria, our energy emissions can be entangled with the electromagnetic waves from biological essences that can have a positive effect on our health and wellness. Moreover, we can amplify the effect of the essences by combining multiple essences with similar frequencies; combine this with sacred geometry or Phi proportions, and you have a powerful formula for energy medicine or as we like to call it quantum nutrition with the potency and effectiveness to help us all be the brightest and best versions of ourselves.

At five years old, I knew I wanted to be a chemical engineer and an Olympian. These two things that seemed so random and unrelated merged beautifully to be the foundation of what I do now. Through an unconventional childhood, I learned to connect with nature, which became the underlying theme for everything in my life. Through professional athletics, I learned the importance of the energy around me, matching the frequency of the reality I wanted and nurturing my energy body. Through engineering, I learned the hard sciences, research and how to harness nature's energy in a way that would have the biggest and best impact on my physical performances and my life. Now, I want to help others do the same.

This has been my path, my story. I recognize everybody has their own unique path and story to tell. I look forward to hearing those stories as we all continue to strive towards living fully, loving wastefully and being passionate ... being a force in all we do!

Chapter 10
Seed for Thought

A couple of years before I retired from mogul skiing at the World Cup in Inawashiro, Japan, I decided to shake things up and change the status quo. The friendly Japanese audience at the event had helped me reach a more peaceful, balanced state of mind than what I normally experience on competition day. The weather was beautiful, and the course was in good condition. On that day, I decided to reward the crowd with a sign of what was to come in the future. Instead of skiing my standard run with the standard air package, I launched a 720 off the bottom air, stuck the landing and skied across the finish line. The run put me on the podium, and the jump surprised everyone except my husband. It was the first 720 ever performed by a female skier in World Cup competition. Sometimes the experiment works, and, when it does, the mindset changes.

Energy ~ Information ~ Consciousness

21st-century science is proving things that mystics, yogis and healers have been talking about for millennia. To explain this in terms of physics let's look again at the famous equation Einstein gave us over 100 years ago:

$$E=mc^2$$

where E = energy, m = mass, c = speed of light.

Have you ever thought about what that really means? Through the conversion factor of the speed of light squared, mass is energy and energy is mass:

Mass = Energy; Energy = Mass

Take your physical mass and multiply it by the conversion factor of the speed of light squared (89,875,517,873,681,800 meters per second) and you get your energy body! Your energy body is far bigger than your physical body. The physical mass we perceive is just energy slowed down enough for our senses to perceive it.

According to Dr. Emoto's work, words have energy, thus:

Energy ~ Words or Information

Dr. William Tiller's research supports the thesis that thoughts have energy. Recall that Jim Self suggests that thought occurs internally with an electric charge, and emotion is tied to the thought in the same manner as a magnetic field vibrates around an electric current. We know through Maxwell's equation that energy is a function of electromagnetic and electrical energy, therefore:

Energy ~ Words or Information ~ Thought (electric) ~ Emotion (magnetic)

I always thought it strange that Einstein picked "c" to represent the speed of light. It could have been used for the obvious term "constant" or for "celeritas," the Latin word for speed, but these choices are not consistent with the other terms and letters used in the equation. Einstein was German, and the German word for light is "licht." Why not L, LS, lg or a lightning bolt symbol? Might it be that what he really meant is energy equals the mass of a human being times consciousness squared?

The definition of consciousness in the Merriam-Webster Dictionary is the quality or state of being aware especially of something within oneself. Could it be that thought is consciousness, and higher levels of consciousness are more intense, clear positive thought tied to more intense clear positive emotions?

According to the Journal of Nature and New Science, consciousness unites all physics. This is the unified field equation. Einstein said energy and information must be interchangeable ~ there must be an equation that relates energy and information. Could it be consciousness or liberated thought?

For the past 400 years, human consciousness and intent have been rejected as possible and significant experimental variables in physical reality. However, today, from the experimental findings of Dr. William Tiller and others, this assumption is no longer true. These and other scientific discoveries indicate that our world has entered a new era.

A Mayan priest from Guatemala shared with me that, we as a society have been in the mud era for thousands of years, i.e., heavy, systematical, fighting, ego and very Newtonian in thought. But now we are in the new light era, 1,000 years of peace, prosperity, love, light and Einsteinian or alchemical principles, where everything is energy and consciousness unites all things. In this era, we recognize the energy in the botanical world and utilize it as the energy medicine of the future, therefore, helping people do amazing things. In this era, we open our minds to the new physics to solve the critical problems of the 21st century.

I am working on a new project where, hopefully,

"The day science begins to study nonphysical phenomena, it will make more progress in one decade than in all the previous centuries of its existence."
Nikola Tesla

we should be able to show through an actual physical transformation that energy is not just felt and witnessed in physical performance but can cause an actual physical transformation of mass through the techniques mentioned here. Change is on the horizon, and I look forward to sharing what I find with you. I also look forward to hearing your stories and experiences. Until then, be a force!

Acknowledgments

My heartfelt thanks goes out to contributors and supporters of my research from the Colorado School of Mines: Professor Persichetti, Chemical Engineering Department; Dr. Hudson, Humanities, Arts & Social Sciences, a huge supporter in "thinking outside the box" of a traditional Mines education heavily focused on the oil and gas industry; Dr. Wu, Head of the Chemistry Department; and Dr. Ganley, Chemical Engineering Department. I owe an extra special round of "snaps and claps" to Professor Persichetti for stepping out of the traditional academic line of thinking to embrace and encourage my nontraditional, more colorful and aromatic research ideas. We need more people like him and Dr. Hudson who understand that education and advancement in medicine and technology are more likely to grow from a bridge across departments than from divided silos of expertise.

To my colleagues and teammates at Phia Alchemy and Phia Lab, thank you for helping create the environment for me to continue this work and all your support along the way. And, most importantly, thanks to my family. To my three-year-old daughter who continues to teach and challenge me in new ways. To my seven-year-old son who is a bright light of abundant energy. And to my love who provides a solid foundation for me to continue to build and pursue my passions … who has always held the string, so the kite can fly.